PENGUIN BOOKS

BOTS

Andrew Leonard is the technology correspondent for the online magazine *Salon* and is a contributing writer for *Wired*
magazine. He has written about technology for a variety of
publications, including *Rolling Stone*, *The Nation*, *The Far
Eastern Economic Review*, *The New York Times Book Review*,
British Esquire, and *San Francisco Magazine*. He lives in Berkeley with his wife Jeni, sister Amy, daughter Tiana, son Eli, dog
Fuzzy, and cat Colette.

BOTS

THE ORIGIN OF NEW SPECIES

Andrew Leonard

PENGUIN BOOKS

PENGUIN BOOKS
Published by the Penguin Group
Penguin Putnam Inc., 375 Hudson Street, New York, New York 10014, U.S.A.
Penguin Books Ltd, 27 Wrights Lane, London W8 5TZ, England
Penguin Books Australia Ltd, Ringwood, Victoria, Australia
Penguin Books Canada Ltd, 10 Alcorn Avenue,
Toronto, Ontario, Canada M4V 3B2
Penguin Books (N.Z.) Ltd, 182–190 Wairau Road, Auckland 10, New Zealand
Penguin India, 210 Chiranjiv Tower, 43 Nehru Place, New Delhi, India, 11009

Penguin Books Ltd, Registered Offices:
Harmondsworth, Middlesex, England

First published in the United States of America by HardWired, 1997
Reprinted by arrangement with Wired Books, Inc.
Published in Penguin Books 1998

10 9 8 7 6 5 4 3 2 1

THE LIBRARY OF CONGRESS HAS CATALOGUED THE HARDCOVER AS FOLLOWS:
Leonard, Andrew, date.
Bots: the origin of new species /Andrew Leonared.—1st ed.
p.   cm.
Includes index.
ISBN 1-888869-05-4 (hc.)
ISBN 0 14 02.7566 5 (pbk.)
1. Intelligent agents (Computer software).   2. Human-computer
interaction.   I. Title.
QA76.76.I58L46   1997
06.3—dc21   97–9274

Printed in the United States of America
Set in Berkeley Book
Designed by Helene Wald Berinsky

For Jeni

CONTENTS

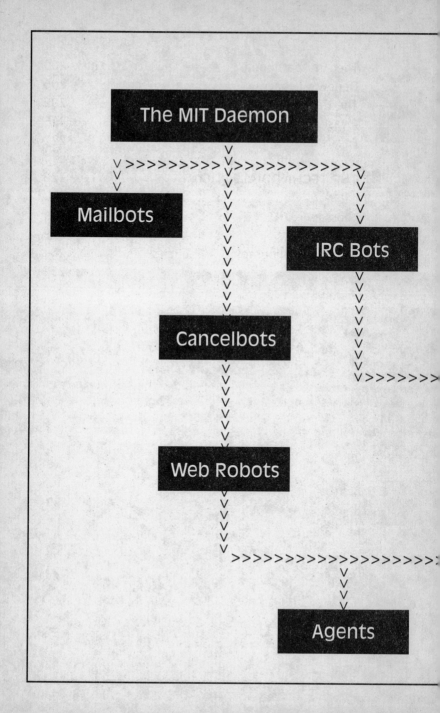

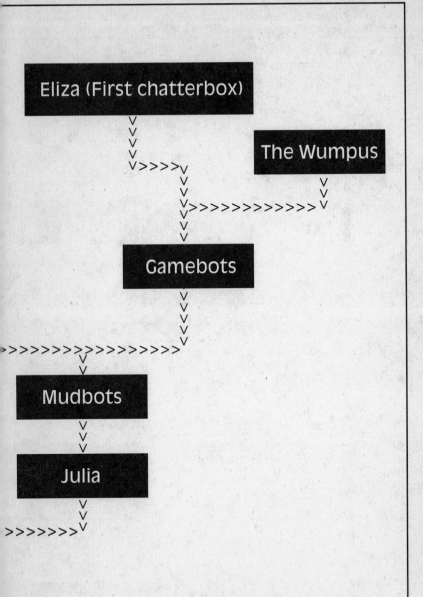

THE BOT FAMILY TREE

ACKNOWLEDGMENTS

Without the patience, attention to detail, and generally relentless competence of my editor, Connie Hale, I would still be floundering in a book-writing fog today. Connie brought order to chaos, and made this book happen. All writers should be so lucky as to have her as their editor.

Connie wasn't the only excellent editor, however, to play a role in the transformation of a column into a magazine story into a book. Miriam Wolf at the *San Francisco Bay Guardian,* and Mark Frauenfelder, Martha Baer, and Kevin Kelly, all at *Wired* magazine, provided ample encouragement and crucial advice. I will never forget that particularly incendiary moment when Kevin Kelly told me over the phone to "let myself go" and strive to be "the Darwin of bots." Dangerous words for a writer to hear, but apparently he knew what he was doing.

My agent Karen Nazor and my publisher Peter Rutten took a chance on an unproven magazine writer. Thanks. Thanks also to the talented crew at HardWired who took my raw text and made it into something really neat—production director Donna Linden, design director Susanna Dulkinys, designer Katja Grubitzsch, and my most scrupulous of copy editors, Rosemary Sheffield.

There are too many friends to thank, but two men, Scott Rosenberg and Alex Cohen, actually read the book and helped steer it toward its final destination. Their input was invaluable, but of course, any mistakes found here are all mine. Finally, there are the three women who share my house, my wife Jeni, sister Amy, and daughter Tiana. They are to be commended with the highest praise for handling the endless fallout from my bot obsession—with grace, forbearance, and good humor. I'd like to promise that I'll never put them through anything like this again, but that, perhaps, wouldn't be prudent.

Andrew Leonard
Berkeley, California

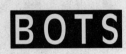

BOTS

1

A PLAGUE OF BARNEYS

THE BARNEY PROBLEM

No one could ever accuse the wizards of Point MOOt of thinking too small.

They dreamed of simulating a fully functional society, lodged within the digital recesses of a computer, true in all essential categories to the principles of the real world. They called this dream "reality modeling."

But just as dreams are never more than a half step from nightmares, the "reality" of Point MOOt never quite "modeled" as expected. Consider, for example, how the wizards endeavored to save their infant civilization from class warfare and ended up with a plague of Barneys.

Back in the spring of 1993, optimism abounded in the Advanced Communication Technologies Laboratory at the University of Texas at Austin. Anything was possible at the ACTLab, a hothouse for experiments in digital culture directed by Professor Allucquére Rosanne Stone. A transsexual specialist in the mysteries of identity and gender bending, Stone encouraged creative approaches to the investigation of social realities. The transformative possibilities of imaginatively constructed online communities particularly fascinated her. Designating a computer to be at the disposal of the Point MOOt wizards—a hodgepodge of undergrads, doctoral students, and intrigued volunteers—she gave them carte blanche to create a brand-new world.

Programmed from a hard-drive tabula rasa, Point MOOt began as a top-down exercise in social engineering. But the wizards anticipated no difficulties in attracting a citizenry. Build it, they reasoned, and the masses would come. Anyone who could connect, via the Internet, to the computer that housed the code for Point MOOt qualified as a potential immigrant. If you could log on, you could move in.

The wizards labored day and night encoding an impressive civic infrastructure. Located at a fictional West Texas crossroads, Point MOOt boasted a university, several strip malls, and a TV station. Cafes, bars, and restaurants lined the streets. Trailer parks and tenement buildings, hospitals and churches, Point MOOt had everything a healthy city requires. There was even an underground cult that worshiped Cthulu, a slime-covered grayish green monster who periodically stormed through town, trampling buildings and kidnapping unlucky citizens. No town is perfect.

Gathered together from construction sites scattered throughout cyberspace, the Point MOOt wizards did not lack for experience. Similar computer-based virtual worlds—known generically as MUDs, or multiuser domains—began appearing in 1979. From their earliest incarnations as online versions of popular fantasy role-playing games such as *Dungeons & Dragons*, MUDs gradually evolved into community hangouts: havens in the digital ether where like-minded individuals could engage in no-holds-barred, text-based "conversation."

Today MUDs have morphed into a dazzling array of forms that include interactive, three-dimensional, animated environments. In a state-of-the-art virtual world, inhabitants can choose full-bodied *avatars* as representations of themselves and navigate through stimuli-packed explosions of sound, pictures, and even video. But as recently as 1993, most communication and experience on the Net were manifested through the vehicle of text. MUDs were worlds built out of words. A MUD visitor typed the command >@look< and saw a text description of a person, place, or thing. MUD objects, rooms, and characters could have various properties—the ability to move, to change, to speak—depending on how the wizards programmed them. But the essential quality of MUDlife was textual.

Point MOOt belonged to a subset of MUDs known as MOOs. The acronym "MOO," short for "MUD object-oriented," referred to the difference between the programming languages used to write the underlying architecture of MUDs and MOOs. The MOO language, developed by Pavel Curtis at the legendary Xerox PARC (Palo Alto Research Center), took the basic ideas of the MUD language to a more accessible level. Many of Point MOOt's wizards—crafty geeks sporting gaudy noms de cyberplume like Chiphead and Chivato, Smack and Racer-X, Warhol and Ogre—were veterans of landmark MOOs, such as Curtis's LambdaMOO or the rollicking post-structuralist playground PMC (Post Modern Culture) MOO. Point MOOt offered these wizards a blank canvas on which to sharpen their skills and to exercise their imaginations.

Though the wizards had differing motivations—some were radical left-wing revolutionaries, some were troublemakers, and some just wanted to build a city—they were united by a desire to make Point MOOt a special place, a new stage in the evolution of MUDs. Their ambitions knew no limits. "We tested both the boundaries of the virtual society and of 'real' society itself," wrote archwizard Allan Alford.

But the wizards soon discovered that testing boundaries is dangerous business. Social engineering is far from a perfect science. Reality is a mess, and reality modeling is even messier. Every problem solved engendered new, unforeseen dilemmas. Every effort at programming a new world led directly to old-world bugs. And just as was true in meatspace—where real people met face-to-face—the business of economics proved to be the toughest nut to crack.

Point MOOt had an economy based on *quota*, or *MOOlah*, as it was sometimes called. If you wanted to build yourself a home, you needed MOOlah. If you wanted to buy products or services offered by other citizens, you had to hand over some MOOlah. MOOlah was both the currency of the realm and the raw matter of construction. One built one's private palace out of MOOlah.

In Point MOOt, MOOlah did not grow on trees, nor did it miraculously materialize in plain sight out in the streets, waiting to be picked up and pocketed, as was commonplace in other MUDs. MOOlah had to be earned. If you wanted to be a productive citizen of Point MOOt, able to enjoy the finer things in life, you had to get a job. Sure, you could always head on over to Nurlene Moot's office at City Hall and ask the suspiciously narrow-minded bureaucrat for a welfare application. But that didn't appeal to everyone. In West Texas, throwing yourself at the mercy of the state wasn't a popular option. Far better that you made the trip to see Buford Moot. Though Buford, like Nurlene, occasionally came off as just another obtuse paper pusher, in his capacity as city planner he could write out job contracts for a long list of available employment opportunities. Most involved working on the city itself: constructing housing, building roads, creating civic spaces. Some were less conventional. The wizards were not afraid to embrace the dark side in their reality modeling. One of Buford's job contracts offered MOOlah to anyone who could concoct a business scam designed to rip off fellow residents.

There was just one catch. To accomplish any of these jobs, you needed to be conversant in the MOO programming language that was to Point MOOt what drywall and concrete are to real-world metropolises. You needed to be MOO-literate.

As programming languages go, MOO isn't especially difficult. It's not as if would-be Point MOOt wage earners were being asked to get down and dirty with machine language digital instructions meant to be read directly by a computer. The MOO language is easy to master; introductory manuals are available free for downloading at locations scattered across the Internet. Still, to many people, the prospect of grappling with *any* programming language at all presents a serious obstacle. Basic programming cannot compare to the cognitively oppressive challenge of, say, abstract math or Heideggerian philosophy, but to a substantial sector of the population the act of programming seems impossibly difficult. Most people would rather not even try.

So the wizards of Point MOOt had a problem. Reality modeling was a noble goal, but requiring people to work at jobs for which they did not have the necessary skills struck some city planners as a tad unfair. Worse, it could prove positively damaging to the fabric of a civic society. The last thing the wizards wanted to see was the immediate stratification of the citizenry of their spanking-new city into upper and lower classes based upon who had the chops to whip up a simple MOO program. The MOO-slinger digerati would sip martinis in expensive cafes, while the programming lumpen waited for welfare checks and sweated in the projects on the outskirts of town.

Rather than stock up on barricades and barbed wire for the inevitable bloody revolution, Point MOOt's wizards sought another way. There was little point in their grand experiment if they let it devolve into sullen chaos right from the inception. There was only one permissible answer: widespread access to easy jobs, jobs that anybody could handle, whether or not they knew how to hack MOO code.

In a private email discussion group used to exchange programming tips and good old gossip, the wizards bandied about possible solutions. And fortunately, as programmers, they were blessed with the power to make real changes. As the ultimate arbiters of every facet of Point MOOt life, the wizards could choose from a considerably wider range of options than your average real-world commissar or ruling party policy wonk. The solution they hit upon proved both imaginative and irresistible. In Point MOOt, no citizen would be forced into penury. If welfare was too shameful and regular work too hard, one choice still remained.

Unemployed? Go kill a Barney.

< >

Deep beneath the city of Point MOOt lay a labyrinth of interconnected subterranean rooms. Inaccessible to most of Point MOOt's citizens, these rooms contained a bizarre assortment of wizard works in progress. In one such room lurked an exceedingly odd contraption—the Barney-spewing machine. Every so often, the Barney-spewing machine lurched into motion and spat out a man dressed in a purple dinosaur suit. A Barney. A digitally constructed clone of that sweeter-than-sweet children's television character who is adored by two-year-olds everywhere, tolerated by their parents, and loathed by all other members of civilized society. Shortly after conception, the Barney, like some hideous beast from the netherworld seeking the light of day, ascended a twisted network of hidden tunnels until it reached the aboveground town of Point MOOt.

To this day, the wizards of Point MOOt hedge and start to mumble when asked why they chose to inflict the likeness of Barney on their fair city. At first, it was just a lark, a snug fit with Point MOOt's Texan/sci-fi theme. In any event, such displays of self-mocking silliness are not all that unusual in online environments: a fair percentage of cyberculture's architects spent far too many hours of their youth watching Saturday morning cartoons.

But origin question notwithstanding, Barneys became a staple of Point MOOt life. There was no avoiding them. They could wander the length and breadth of the MOO. They could open the doors to any room. And they were vocal to the point of becoming a public nuisance.

Allan Alford, a fast-talking Texan and the single most important driving force behind the creation and evolution of Point MOOt, did not personally initiate the Barney phenomenon. But he remembers it well, with a wistful nostalgia in wry counterpoint to his otherwise sincere, and serious, ruminations on the progressive social goals of Point MOOt.

"They roamed the MOO singing the Barney song," says Alford. (A terror indeed, as anyone who has been exposed to the song is well aware.)

Woe betide the citizen unable to keep cool in the face of a serenading Barney. Physical violence against a Barney (expressed in MOOish by text commands such as >@hit Barney< or >@kick Barney<) was not the answer. If attacked, Barneys frequently fell apart—a lopped-off arm here, an amputated leg there—then regenerated, one new Barney from each separate body part. The reproductive feat ensured a steady Barney population explosion, since the human residents of Point MOOt often found it difficult to restrain their anti-Barney animosities.

The Barneys were pests.

"They were actively annoying—free-roving, self-spawning, and could pick locks. That's what made them horrible," says Alford.

The wizards decided to kill two MOObirds with one MOOstone. They had too many Barneys and not enough jobs. So they created a new job: Barney hunter, a job for the masses, requiring of citizens only that they be able to type the command >@shoot Barney<.

You had to be equipped with a Barney Blaster gun before you could start taking potshots, but that was easy enough. Harley Moot, Point MOOt's chief bounty hunter, had access to an inexhaustible supply of Barney Blasters, available to the general public without waiting periods or credit checks. Point the Barney Blaster at a Barney, type >@shoot Barney<, and boom! A purple man-size dinosaur hit the turf, shredding into several nonregenerating parts. Harley Moot paid one unit of MOOlah for each dead Barney, payable upon delivery of Barney's severed head.

For a brief moment, the wizards thought their problem solved. The techno-proletariat had been assured a minimum wage. Barney hunting became Point MOOt's most popular pastime. A steady stream of fresh Point MOOt immigrants lined up at Harley Moot's door.

But the central importance of Barney hunting to the Point MOOt economy soon led to unexpected results. All of Point MOOt's citizens, even the ones who did know their way through mazes of MOO code, started to pay close attention to the purple man-beast with the silly smile. They discovered that if they typed in different MOO commands besides >@shoot< (such as >@feed< or >@impregnate<) they could cause interesting things to happen.

Impregnated Barneys gave birth to new Barneys, saddled at birth with individual names that combined the name "Barney" with the online nickname of the human who had forced his or her attention on its parent. (If you saw a Chivato-Barney wandering around, you knew Chivato had been naughty.) Barneys stuffed with food presented a worse problem. Barneys were insatiable: they continued eating nonstop until ballooning in size and exploding. And the fragments of an exploded Barney spawned five new Barneys! One mad Point MOOt scientist figured out a simple series of MOO commands that looped the feeding cycle. After the first overfed Barney exploded and reproduced, all the new Barneys commenced eating until oversatiation. Over and over again.

The combined production of the Barney-spewing machine, Barneys looped into a feeding frenzy, and the normal population growth attributable to kicked and beaten Barneys ensured that the wizards of Point MOOt, after narrowly averting the tragedy of class warfare, now had a new problem. The Barney problem.

At any given moment, any location in Point MOOt could be overrun by hundreds of vacuously grinning purple dinosaurs, all singing the Barney song. "They would come and go in huge waves," says Alford. "And they far exceeded the number of Barney hunters."

From the perspective of a Point MOOt citizen, whose view of the world was confined to text scrolling across a computer monitor, the experience of a Barney onslaught was disorienting, even frightening. Suddenly the monitor exploded with an endless stream of "I love you, you love me" sentences spiraling down the screen. The Barney scourge put a real damper on the possibilities of civic life in Point MOOt.

A CERTAIN QUALITY OF INHUMANITY

Barney was a bot. Not a very smart bot, and not a very useful bot, but a bot nonetheless.

A bot is a software version of a mechanical robot. Like a mechanical robot, it is guided by algorithmic rules of behavior—*if this happens, do that; if that happens, do this.* But instead of clanking around a laboratory room bumping into walls, software robots are programs that maneuver through cyberspace, bouncing off of communication protocols and operating systems.

Strings of code written by everyone from teenage chat room lurkers to topflight computer scientists, bots are variously designed to carry on conversations, act as human surrogates, or achieve specific tasks—in particular, to seek out and retrieve information. Bots entertain, annoy, work, and play.

Over the past decade, as the online universe has exploded supernova-like into every interstice of modern life, bots have begun to flourish. In every neighborhood of the Net, bots either lurk behind the scenes or demand attention, front and center. Mailbots filter electronic mail, preventing junk mail and *spam* advertising from clogging up our online mailboxes. Chatterbots carry on whimsical conversations in online, real-time text environments, such as chat rooms or MUDs. Cancelbots seek out unwanted expression and erase it from electronic bulletin boards. Gamebots populate computer game environments with believable characters and wily foes. Web robots explore the hyperlinked reality of cyberspace, mapping out and indexing the vast quantities of information available through the World Wide Web.

< >

Bots are the first indigenous species of cyberspace, a class of creatures dazzling in its infinite variety. Web robots, spiders, wanderers, and worms. Cancelbots, modbots, Lazarus, and the Automoose. Softbots, userbots, taskbots, chatterbots, knowbots, and mailbots. MrBot and MrsBot. Bartender-bots, BalooBear bots, and bolo bots. Warbots, clonebots, crashbots, floodbots, annoybots, hackbots, and Vladbots. Turing bots. Tsunami bots. Gaybots, gossipbots, and gamebots. Prostibots. Conceptbots and RoverBots. Skeletonbots, spybots, slothbots, and spambots. Xbots and metabots. Eggdrop bots. Motorcycle-bull-dyke bots.

Most bots, unlike the Barney bots, refrain from singing silly songs or exploding into overeating-induced spasms of bodily fragmentation, but all bots share some aspects of the Barney experience. Their evolution is full of high hopes, unexpected wrong turns, near catastrophes, and wild exuberance.

In one form or another, bots have been around since the early 1960s. But there is no consensus on what particular sequence of encoded ones and zeros truly classifies a bot. Bot genetic structures remain inadequately mapped. The word *bot*—a slang truncation of *robot*—describes everything from a simple logon script (like one that might save a user the trouble of typing in a phone number, a password, and a user identification code every time that user wants to go online) to complex programs written in the latest, most-advanced programming languages and designed to execute tasks that most humans would find impossible.

At Point MOOt, Barney was just one of many bots proud to

call virtual West Texas their home. In the early summer of 1994, fully half of Point MOOt's population were bots. Nurlene, Buford, and Harley Moot weren't just closely related. In addition to sharing a last name, they shared a certain quality of inhumanity. They were bots.

As were all the city officials. Bots greased the wheels of Point MOOt bureaucracy. They were an essential part of the reality modeling process. In Point MOOt, you never knew where the next bot would turn up. Point MOOt had bum bots, ne'er-do-well homeless wanderers who shuffled up to citizens, panhandled them, and refused to leave unless given some MOOlah. Annoying, often, but not half so bad as the hooker bots, who evinced no shame whatsoever about propositioning citizens in broad daylight. One brazen hussy hooker bot once strolled right into Buford Moot's office while he was interviewing a prospective job applicant! Point MOOt even had a pseudo-Hunter Thompson bot—a beer-swilling gonzo journalist who hung around in cafes and bars and had an unpleasant habit of suddenly losing his cool, leaping up, and spraying Mace at the nearest human.

Allan Alford's personal favorite was Chico, a bot formed in the persona of a foreign exchange student. In this guise, the bot could get away with making syntactical errors that would otherwise betray his robotic DNA. Alford toiled for months on Chico, writing elaborate sentence parsing algorithms designed to enable the bot to learn new vocabulary and master complicated grammar.

The bots added color to Point MOOt. More important, though, the city official bots saved the wizards from having to deal with the hassles and minutiae of running the city on a daily basis.

"My first idea," says Alford, "was to hire people to become bureaucrats and pay them with quota [MOOlah]. But no one wanted to do it, or if they did want to do it, they weren't around when they were needed. . . . But then I suddenly realized—this is a programmed environment. Bots were the key. I began developing bots to essentially become town bureaucrats. . . .

"Nurlene would talk to you, interview you, explain the rules, make you sign the right forms, and then transfer you to the welfare system," says Alford.

Barney started as a joke and evolved into an economic cog. But Buford and Nurlene were designed from the beginning as labor-saving devices. As such, they unconsciously harked back to the derivation of the word *robot*. First used by Czechoslovakian science fiction writer Karel Čapek to describe mechanical beings, the word means "forced labor" in the original Czech.

Buford and Nurlene were tools. Slaves, even, or at best, indentured servants. The bum bots and hooker bots were characters, role players in the community. Barney ended up being a little of both. A bot can be many things: it's a concept that defies easy categorization.

< >

"*Bot* is short for *robot,* which is cooler than *program,*" says John Leth-Nissen, a hacker who hangs out on the Internet Relay Chat (IRC) network. And that captures part of the bot equation. Bots are cool. They stoke our imaginations with the promise of a universe populated by beings other than ourselves, beings that can surprise us, beings that are both our servants and, possibly, our enemies. Bots, which are here, now, and growing in number and power every day, are advance scouts from the future.

But the gee-whiz cool factor only scratches the surface of why bot movements should be watched with an eagle eye. Bots represent both the aspirations of the cutting edge of the computer research community and the voracious hopes and dreams of software entrepreneurs. Bots are a test bed for experiments in the arena of artificial intelligence, and they're a stab at solving the ever sticky problem of making the interface between humans and computers effortless and enjoyable. Bots are a way of thinking about how we interact with computers: our desire to attribute personality to programs is as much about our desire to fill the world with myths and legends as it is about any definable, hard-coded reality.

Some of the rhetoric about bots aims too high. Sherry Turkle, an MIT psychologist who specializes in online interactions, defines a bot as a "small artificially intelligent program." She may be jumping the gun. Artificial intelligence is far from a done deal. The concept of intelligence, in daily life, resists easy definition; to attribute it to a computer program is to beg for debate. Is intelligence the ability to speak a natural language? to have common sense? to reason? Or is it, as Alan Turing suggested, a computer program's ability to pass itself off as a human being? Is it merely a matter of tricks and prestidigitation?

Certainly, one would be hard put to make the case that the Barney bot had much more intelligence than a doorknob. On the chain of bot-being, the Barney bot ranks pretty low. But then again, if one could suspend disbelief long enough to accept that men in purple dinosaur suits inhabited the town of Point MOOt, perhaps one could also succeed in pretending that the Barneys displayed a modicum of intelligence.

A more accurate definition than Turkle's might run as follows: a bot is a *supposedly intelligent* software program that is *autonomous,* is endowed with *personality,* and usually, but not always, performs a *service.*

Personality implies that the bot displays some aspect of human behavior or has in some way been anthropomorphized.

Barney had personality galore, even if most of his acquaintances found it irredeemably repulsive. *Autonomous* means that the bot must be able to do its work without direct human supervision. Barney was autonomous: he wandered from room to room in Point MOOt without being told where to go in advance or having to be prodded at each step. And Barney performed a *service:* originally he provided comic relief; later he became an economic mainstay.

The service aspect of bots is crucial, a character trait running at cross angles to issues of autonomy, personality, or intelligence. Nurlene Moot, the city official, had a purpose in life—not just as the administrator of the dole but also as interface between the wizards and their creation. Through Nurlene the wizards ensured that certain tasks were carried out that the wizards either refused to do or were too busy to do. Nurlene also served as interface between the citizens of Point MOOt and the interior workings of the city.

The service/interface aspect is what makes a bot something greater than a curiosity. Bots are the first precursors to the *intelligent agents* that many visionaries see as indispensable companions to humans in the not-too-distant future. Intelligent agents are software programs designed to help human beings deal with the overwhelming information overload that is the most obvious drawback to the information age.

Though agents and bots significantly overlap, the two categories are not exactly equivalent. Agents do not require the accoutrements of personality: human names, the ability to crack a bad joke, zany habits. And though bots can be very useful, they are not forced, like agents, to work hard for their digital room and board. A chatterbot that sits in a MUD and talks about ice hockey with anyone who says hello is not an agent.

But for an agent to be really successful in its chosen task, it may need to be a bot or, as some in the industry prefer to say, a *believable agent*. True botness makes a computer program or a computer environment more approachable, more entertaining, more user-friendly—drastically important considerations for those who wish to create successful agent prototypes, either for academic research or for the consumer marketplace.

Listen for a moment to one of the leading apostles of agent religion, Nicholas Negroponte, director of MIT's Media Lab, as he explains why agents should be considered an attractive necessity. After discussing the baffling responsibilities of hectic modern life, he wonders, "Wouldn't you really prefer to run your home and office life with a gaggle of well-trained butlers (to answer the telephone), maids (to make the hospital corners [on your bed]), secretaries (to filter the world), accountants or brokers (to manage your money), and on some occasions, cooks, gardeners, and chauffeurs when there were too many guests, weeds, or cars on the road?"

In Negroponte's vision, these tasks are performed not by actual humans but by individualized, personified computer programs. Negroponte calls his digital butlers "agents." But he could just as well call them bots. *Bot: Find me the best price on that CD, get flowers for my mom, keep me posted on the latest developments in Mozambique.*

Negroponte's vision is shared by a horde of entrepreneurs currently stampeding into cyberspace with agent products. The giants of the computer industry—Microsoft and Intel, IBM and Apple—are funneling millions of dollars into agent research. But both the ivory-tower academics and the corporate researchers are pushing down from the top, seeking new tools to aid them in their quest for the secrets of intelligence or bottom-line profit bonanzas.

Bots, meanwhile, just happen, out there in the vast and growing wildernesses of the Net. Instead of seeking the bottom line, they are growing from the bottom up, a grassroots phenomenon, as likely to be conceived by bored teenagers looking for fun as by PhDs from Stanford or MIT. There are no international conferences devoted to bots, no high-powered academic departments with "autonomous bot" groups. The evolution of the Barney bots of Point MOOt was accidental, unscripted, and unpredictable. And there, in a nutshell, is the story of the evolution of all bots. They are the spawn of the Net's anarchy and decentralization, the product of a thousand different hackers writing code on a thousand different computers.

Bots, like all creatures, belong to their environment. And in this case, that environment is the Net. The proliferating MUDs, the thousands of bulletin boards that make up Usenet news, the practically infinite chat rooms of the IRC network, the inexorably spreading and morphing World Wide Web—only in the most accidental of ways can the Net be said to have been planned or organized or regimented. It is a massively successful example of the power of parallel distribution—endlessly inventive, endlessly changing.

This multiplicity of Net environments is one reason why attempting to discern a particular genealogy or anatomy of bots is a mission fraught with semantic and logistic peril. Bots comprise not just one new species but a complete spectrum of new species, a brand-new phylum under the digital sun.

< >

As interface, as artificial intelligence experiment, as pure playthings, bots exert a strange attraction on the human inhabitants of cyberspace. The bots of Point MOOt, and the Net, make for a good story. But so do the bot problems. The evidence of a plague of Barneys is a warning bell signaling more than just what can go wrong in the case of improper or incompletely thought-out programming.

Bots don't have to be benign, and bot misbehavior doesn't have to be accidental. Bots can be instructed to do whatever their creators want them to do, which means that along with their potential to do good they can also do a whole lot of evil.

With newbie botmaster wannabes joining the Net in huge numbers every day, bad bot shenanigans are bound to get worse. The Net is no longer a playground for just the technically clued in. It is increasingly complex, and it increasingly reflects the strains and pressures of the real world. Commercial incentives fuel ever growing levels of investment. The steady influx into cyberspace of the age-old viral hatreds and lunacies that infect the world's face-to-face culture will only increase. Bots will be—and are—the vehicle for uncontrollable passions.

Bots aren't just cool. They're trouble.

2

DAEMONS AND DARWIN

THE DAEMON

> They are the envoys and interpreters that ply between heaven and earth, flying upward with our worship and our prayers, and descending with the heavenly answers and commandments, and since they are between the two estates they weld both sides together and merge them into one great whole. They form the medium of the prophetic arts, of the priestly rites of sacrifice, initiation, and incantation, of divination and of sorcery, for the divine will not mingle directly with the human, and it is only through the mediation of the spirit world that man can have any intercourse, whether waking or sleeping, with the gods.
>
> —*The Symposium*, PLATO

Socrates had a bot.

Not in the strictest sense, of course. The word *bot* did not belong to the Socratic vocabulary. But Socrates did have a non-human companion, or so he claimed. He called this companion a *daemon*. Intelligent and always ready to offer good advice, Socrates' daemon could be trusted to act without prompting. These are good bottish qualities all, and the correspondences were not mere coincidental similarities. Real, hard-coded linguistic and symbolic links abound between that daemon and today's bots.

Socrates' daemon, just like a contemporary bot, was a mixed blessing, a source of as much ill fortune as good. In 399 B.C. the petty democrats who ruled Athens indicted Socrates for crimes against the state, on two counts: "corruption of the young" and "believing in deities of his own invention instead of the gods recognized by the state." Socrates' daemon, declared the prosecutor Meletus, was one such deity: a false god, de facto evidence of impiety.

At his trial, Socrates described his daemon as an internal oracle, a kind of divining (and divine) rod that led him down the correct path of speech and action—a turbocharged conscience. Not so, alleged Socrates' accusers. Such waffling, they argued, missed the essential point. Socrates, according to the prosecution, really believed that the daemon was an independent supernatural creature that had chosen to associate itself personally with the philosopher. Not internal but external. Not a conscience but a holy spirit. Or unholy, depending on whose god one chose to believe in.

Both charges against Socrates were bunk. Political rivalries underlay the trial's rhetoric. Socrates had long been associated with the notorious troublemakers Critias and Alcibiades, a pair of political adventurers who represented everything that threatened the democratic clique then in power in Athens.

Show trial it may have been, but show trials are usually lethal. Socrates spurned the opportunity to plea-bargain, accepted the death sentence, and willingly drank the poison hemlock. His reputation, of course, lives on, unsullied by insidious accusations.

His daemon was less fortunate. Up to the point of that trial, the concept of daemon bore with it no inherently malignant meaning. As Plato wrote in *The Symposium,* "the envoys and interpreters that ply between heaven and earth" served as vital role players in the drama of gods and humans, invaluable intermediaries between realms that did not touch. After the trial, the daemonic reputation began a long, downward spiral. In the Manichean world of Judeo-Christian culture, daemons were assigned to the dark side. Daemons became demons, fallen angels, creatures of evil, cloven-footed, fork-tailed inhabitants of Hades, intent on dooming all humanity to eternal perdition.

Ever since Socrates and his impressionable pupils walked the Athenian agora, a steady drumbeat of antidaemonic propaganda has resounded. Yet images persist of the daemon (or demon) as helper, rather than as hurtful fiend. Pre-Christian cosmologies abounded with daemons—sprites, fairies, leprechauns—who occasionally could be counted on for a lending hand or sage word of advice. And despite every effort by Christian authorities to cast each such pagan survivor as a minion of Lucifer, their good reputation could not be entirely besmirched.

If carefully unpacked, the word *daemon,* however spelled, uncovers a provocative and useful dualism. An intermediary with another world doesn't have to be beneficent. Yet neither is it compelled to be nefarious. It can be both, flip-flopping between positive and negative states—depending on context or perception, on the vagaries of politics, or the whims of the fickle masses. The Middle Ages were rife with examples: in times of crisis, a wise woman became a witch, and daemons got the blame.

Daemons are both good *and* evil, a fact that became clearer and clearer as the Western world began to clamber out of the depths of centuries of prejudice and superstition. And nowhere was this more apparent than in the arena of science, where cool-eyed devotees of hard-nosed experimentation and research weren't afraid to consider any alternative as an explanation for events or as a possible catalyst for change. Science has inspired something of a daemonic revival. And no single person has been more responsible for this comeback, or for the eventual daemon/bot synthesis, than nineteenth-century physicist James Clerk Maxwell.

Considered second only to Isaac Newton as one of the primary theoretical architects of modern physics, Maxwell pursued the ultimate scientific dream. He sought the subversion of one of the fundamental premises of his field: the second law of thermodynamics, which holds that the entropy of the universe is always increasing. The second law mandates that in order to get something (such as heat or light), you have to give something (energy). Maxwell planned to get around that restriction through the assistance of a molecular-size intelligent being: Maxwell's Demon.

In Maxwell's proposed experiment, Maxwell's Demon stood next to a sliding door separating two sealed chambers of a box containing free-ranging molecules of gas. Monitoring the speed at which the various molecules bounced around the chambers, the Demon could tell which specific molecules possessed higher or lower energy states. Deftly sliding the door open and shut at the correct moments, he could separate these molecules into two different groups based on relative energy level. The end effect would be to heat one room by gathering high-energy molecules together and to cool the other with the low-energy remainder—all without expending any energy to actually move the individual molecules.

A perpetual source of energy was just one application of a successful Maxwellian Demon. However, as many later observers noted, the hypothesis poses a number of problems, not least of which is "finding a Demon and convincing him of his job," as one Internet-based commentator pointed out. There's also that old bugaboo of quantum physics, the Heisenberg uncertainty principle, which states that one can't know the location and speed of a single molecule at a single moment in time. Just the act of observation, even by a fiendishly small being, alters the properties of the observed object.

Neither inward oracle nor false god, Maxwell's Demon was hardly evil, outside a little chaos that might be caused by a practical method for producing perpetual power. The worst one could say is that the tiny demon represented wishful thinking on a gigantic scale: Wouldn't it be nice to have little helpers doing our bidding, in willing defiance of the laws of space and time? Wouldn't it be grand to extend our limited human abilities into new, otherworldly dimensions?

Still, demons continued to suffer from unrelenting bad press in the world at large. Then came computer science.

FROM SOCRATES TO MIT

In 1958 a researcher at the Massachusetts Institute of Technology's Lincoln Laboratory postulated one of the earliest experimental models of an artificially intelligent computer program. The researcher, Oliver Selfridge, called the proposed program "Pandemonium." Pandemonium, wrote artificial intelligence (AI) historian Daniel Crevier, would "look like Milton's capital of Hell: a screaming chorus of demons, all yelling their wishes to a master decision-making demon." Each yell represented a decision by a demon peon (actually a small subprogram). The master demon, in turn, made its decisions by listening to who shouted loudest and incorporating that information into an overall emergent plan.

For technical reasons, Selfridge's program never was implemented as Pandemonium. But Selfridge's theoretical contribution—the demon that waits for a discrete event and then responds to it, as part of some larger coordinated or parallelly processed strategy—proved immensely influential in several subschools of modern AI, including expert systems and neural nets.

Selfridge's demons had little of the flair of Maxwell's beastie. Not much personality, not much latitude for action except for the occasional yell. But Selfridge, or at least the computer scientists working in Selfridge's wake, had one great advantage over the nineteenth-century physicist. The astonishingly fast growth of computers—in terms of processing power, memory, and ever decreasing cost—facilitated the creation of new worlds and new ideas. Creating a real demon out of earthly clay is impossible. But creating a virtual demon is another story. Computer science's great and tantalizing promise is that it will help us smash all boundaries, giving old dreams new life.

Just a few years after Selfridge dreamed up Pandemonium, and just a few miles down the road, another team of MIT researchers similarly fell under the demonic spell, choosing to cast their research and experimentation in the revealing cloak of metaphor and myth. It was the start of a long love affair between programmers and daemons.

The year was 1963. At the time, the staff, faculty, and students who frequented the Computation Center on the main MIT campus shared access to one computer, an immensely cumbersome, hulking IBM 7094. A monster mainframe, the

7094 sprawled over four thousand square feet and required nearly fifty tons of air-conditioning equipment. All for 32K of memory, hardly enough for a state-of-the-art pocket calculator today. Despite the computer's great size, every second of access time to the machine was precious. The team worked under an express research mandate to seek better, more efficient ways of utilizing the 7094's processing power.

Led by Professor Fernando Corbato, a founding member of the MIT Laboratory for Computer Science, the team had already made one major contribution to the history of computer science by devising the first computer time-sharing scheme, in which computer users at individual terminals could simultaneously run programs on a mainframe computer while logged in from different locations. Time-sharing was a major step forward—decades later, Corbato received the prestigious A. M. Turing Award for his "pioneering work" in developing "multiple access computer systems." But broad-scale initiatives weren't the only focus of Corbato and his graduate students. The team endlessly hunted for new tricks, or *hacks,* that would enable humans to avoid mindless busywork or drudge labor. In doing so, they reflected the prevailing attitude, then as now in the computer science community, that anything that can be automated, so as to save human effort, should be.

Corbato, now MIT's Ford Professor of Engineering, remembers coming up with the idea for one such trick: a process constantly resident in the 7094's operating system that took care of a useful task without requiring constant human supervision. Such a process can best be thought of as a computer program in action, a sequence of coded instructions to a computer launched into digital motion. Corbato's process had one job; it endlessly scanned the 7094's databanks looking for files that had been modified since its last scan.

Backing up new files is a mundane computing chore. On a

mainframe computer used by many people from multiple terminals, it can be a major hassle. But with the help of Corbato's process, the MIT computer scientists no longer had to be bothered by such basic housekeeping demands. Every time it found a modified file, the process saved the file to a tape backup, without requiring direct human authorization. The beauty of it all was that this particular process, once launched, stayed operating as long as the computer itself continued running. The process was autonomous.

Considered a virtuoso programmer by his peers, Corbato came, like many other early computer scientists, from a background in physics research. In homage to Maxwell's Demon, he decided to name his new autonomous helper a demon. Like Maxwell's Demon, Corbato's demon admirably filled the role of laboratory assistant. But instead of shuffling molecules, Corbato's demon sorted computer bits—a far easier task.

Corbato's demon was a neat hack. But to at least one member of the team, researcher Michael Bailey, the prospect of introducing a demon into the innards of the 7094 seemed a bit, well, hellish. The British-born Bailey suggested that the team adopt the British spelling of the word—*daemon*—in a conscious attempt to conjure up the good will of Socrates' beneficent spirit.

When Corbato's team moved on to a new time-sharing project, they took the daemon concept with them, convinced that such spirits could be useful for an endless number of jobs. And they were right. As the narrative of computer history unfolded—from the antediluvian era of the mainframe, through the personal computer revolution, and right up into the age of networked cyberspace—daemons became omnipresent citizens of the silicon universe.

Today daemons are a programmer's best friend, one of the

lubricants that keeps the electronic machinery of bits and bytes moving smoothly. Regular users of email meet up constantly with the vigilant mailer daemon, a vital cog in the electronic post office that deals with a host of mailing problems (messages sent to nonexistent addresses, overflowing mailboxes, forwarding and readdressing needs). An HTTP (hypertext transfer protocol) daemon resides in a computer that stores World Wide Web homepages, always on the alert to respond to requests from other computers asking for access to local Web documents. Printer-spool daemons monitor computer file servers to see if any files are waiting to be sent to the printer. A clock daemon executes instructions at specific times.

These myriad daemons, from the original IBM 7094 daemon on forward, all share some basic characteristics. They are processes that run *in the background*—that is, they aren't obvious to a user who might be logged in. They are invisible in the normal course of events, conscientiously taking care of their business out of sight, out of mind. Daemons are also able to respond to environmental changes, to react to events. And once launched, they are out of direct control—they autonomously complete their tasks.

Autonomy is the crucial variable, the dividing line between dull computer clay and rich digital life. A true daemon can act of its own accord. As such, it is as different from a typical software program (a word processor, say) as a clock is from a hammer. Put the hammer down, and it is useless, a dead object. The clock never stops, as long as it has power.

Today Corbato downplays the significance of the 7094 daemon. Even in 1963 the concept of a process that ran in the background, simultaneously with other processes or inside an operating system, wasn't all that new, he says. The main distinction between daemons and other processes was that daemons were doing their thing while humans were active on the system. But even then, says Corbato, "the only new question was one of labeling."

Whether or not Corbato is being fair to himself in deprecating the importance of the daemon as a technical breakthrough is essentially irrelevant. The symbolic importance of labeling should not be underestimated. When Corbato and his team chose the name *daemon,* they engaged in a daring act. They summoned up a spirit from ancient Greece. They declared that a computer program is more than just a set of instructions, a sequence of algorithms. It is an intermediary, not between heaven and human, but between the digital and the biological.

"The divine will not mingle directly with the human, and it is only through the mediation of the spirit world that man can have any intercourse, whether waking or sleeping, with the gods," wrote Plato. Today computer science has replaced the unknowable domain of the gods with the overwhelming, overpowering domain of unlimited information. The ones and zeros that make up the bricks and masonry of cyberspace are as ungraspable to normal ken as the nectar and ambrosia of Mount Olympus were unattainable to the ancient Achaeans. Without our daemons to intercede for us, we would be lost.

THE UR-BOT

The affinities of all the beings of the same class have sometimes been represented by a great tree. I believe this simile largely speaks the truth. The green and budding twigs may represent existing species; and those produced during former years may represent the long succession of extinct species. At each period of growth all the growing twigs have tried to branch out on all sides, and to overtop and kill the surrounding twigs and branches, in the same manner as species and groups of species have at all times overmastered other species in the great battle for life. . . .

As buds give rise by growth to fresh buds, and these, if vigorous, branch out and overtop on all sides many a feebler branch, so by generation I believe it has been with the great Tree of Life, which fills with its dead and broken branches the crust of the earth, and covers the surface with its ever-branching and beautiful ramifications.

—*The Origin of Species,* CHARLES DARWIN

Corbato's daemon belongs at the top of the great Tree of Bots. Call it the ur-bot, the primeval form to which all present and future bots owe ancestry. Of course, there is no received authority that will substantiate this claim; to the contrary, a legion of programmers will scoff at the assertion. Bots and daemons, they will claim, have little in common, and they will quote chapter and verse from a litany of programming arcana to prove their point. What's more, even if these programmers do concede that certain types of bots—cancelbots and Web robots, for instance—incorporate daemonic-style autonomy and environmental reactivity, they will still dispute the casual lumping together of so many wildly different sequences of code under the single semantic umbrella of bots.

Those who find fault with such a categorization of bots may well be correct. On its surface, the idea of tracing biological lines of descent, or consanguinity, between an assortment of different computer programs is outright nonsense, an exercise in quixotic genealogy, metaphorical madness.

Bots are far from being, as Charles Darwin wrote in *The Origin of Species,* "organic beings in a state of nature." They have no instincts, they do not grow old, they do not bleed. They do not share the same DNA, aside from their common fundamental atomic structure as sequences of ones and zeros. Bots are written in a host of different and genetically incompatible computer languages—Lisp, Pascal, C, C++, Lex, Basic, Logo, Perl, and Java. They are not sexually dimorphic; they do not reproduce. (At least not yet: the most recent advances in genetic algorithm programming promise an imminent world in which computer programs *do* spawn offspring and mutate out in the wild, but such creatures have yet to leave the research laboratory to swim free on the Net.)

The bot family tree is a confused and contradictory plant, a warped and twisted structure as unlike Darwin's great Tree of Life as a blackberry bush is unlike a weeping willow. Yet, despite their inorganic nature, the various bot species are related—by circumstance, by example, and by habitat. Corbato's daemon influenced generations of MIT graduate students who went on to create gamebots and chatterbots that owed, at the very least, a debt of inspiration to the Daemon. Point MOOt's Allan Alford discovered the chatterbot algorithms for his bot Chico in a *Scientific American* article that explained the mechanics of a supposedly artificially intelligent program that had been discovered posting odd messages to a Usenet newsgroup bulletin board. Michael Mauldin, the author of Julia, cyberspace's most famous chatterbot, also wrote the Lycos spider, one of the first Web robots to gain notoriety on the World Wide Web. For at least a brief period during 1996, visitors to AlphaWorld, a state-of-the-art three-dimensional virtual world, could chat with a bartender-bot whose core code was nearly identical to that of the first chatterbot, Eliza, who had been born some three decades earlier.

Many more such linkages can be made. But while simply tracing the connections and seeking out linear relationships is enlightening and entertaining, it misses the point about bots. Those who would pin each separate bot to a piece of cardboard and dissect its algorithms one by one in the hopes of actually discerning a unified theory of bot-hood are misdirecting their energy. All acts of classification, whether of animal, mineral, vegetable, or digital, are arbitrary. Even Darwin, struggling to explain why certain organic beings could be considered species, while others merited only the title "varieties," conceded he made his own distinctions solely "for the sake of convenience."

Classification may be arbitrary, but it is also necessary. We classify in order to understand, posit relationships to illuminate theories, create metaphors in search of symbolic truth. As paleontologist Stephen Jay Gould, one of Darwin's most articulate modern defenders, observed, the art of classification, or taxonomy, is a tool for uncovering meaning: "Classifications are theories about the basis of natural order, not dull catalogues compiled only to avoid chaos."

In the case of bots, we deal with unnatural order, but that makes taking account of each metaphor, each way in which people think of bots, even more important. The common strand that holds all the species of bots together, from the first Daemon to the latest agent to hit the Net, is that each bot stands as a metaphorical bridge between the natural and the unnatural. That humans feel impelled to embody software programs as demons or people or animals—rather than emphasize the trivial technicalities of their architecture—is the sap flowing through this great tree.

And yet these software creatures are actual entities beneath their metaphorical trappings. Even if their physical manifestation is no more than the flicker of electric current through a silicon computer chip, bots—from the Daemon onward—do exist in the real world. And they are changing, adding new features and shedding old bugs, increasing in numbers and penetrating into new environments. To borrow a metaphor from that most potent of all Darwinist claims to truth, they are *evolving*.

THE TECHNODIALECTIC

Evolution, for Darwin, can be summed up as the survival of the fittest through natural selection, or more crudely, as the subtitle of *The Origin of Species* proclaims, "the preservation of favored races in the struggle for life." There have been numerous modifications, additions, and refutations of Darwinism in the 130 years since the publication of *The Origin of Species,* but the metaphorical power of the original theory still holds strong.

The key variable, in terms of determining fitness, is habitat. The better a species is fitted to its habitat, the more likely it is to prosper and reproduce.

The bot habitat is cyberspace. For the first bots, which included Corbato's daemon and other early bot prototypes, cyberspace was a limited domain—the operating system of a single computer. For today's bots, cyberspace is unlimited in extent. It is the sum total of all the interconnected computers in the world—the ever growing, ceaselessly squirming, restlessly twisting Internet, as fertile today as the primeval Earth.

Today's Net consists of a myriad of new environments, each with its own indigenous species. Instead of climatic differences, there are changes in programming languages. Instead of natural selection, software upgrades doom or bless. Usenet newsgroups, IRC rooms, the countless subvarieties of the MUD world (TinyMUDs and UnterMUDs, AberMUDs and LambdaMOOs, WOOs, MUSHes, MUCKs)—each has its own predators, quicksand and avalanche threats, and meddling natives.

Darwin theorized that one is more likely to discover a greater variety of species in a large land mass or oceanic environment than in a small one. Enhanced competitive pressure and ecological heterogeneity combined to promote the evolution, through natural selection, of relatively more numerous species. And certainly that is true of the Net. The Net, which is both an entire ecology of its own and a set of interconnected ecologies (Usenet, IRC, the Web, MUDs), has encouraged a bot diaspora. And just as certain plants developed attractive flowers to lure pollinating insects, so too have bots evolved to take advantage of the peculiarities of their surroundings. A Web robot must be able to understand, or parse, the language of the Web, HTML. A cancelbot needs to know just how to interpret the header that each bulletin board message carries to identify itself to inquisitive computers. A bot designed to work in a TinyMUD may not work in a MOO, because of the differences in programming languages used to construct TinyMUDs and MOOs. As these various environments have become more complex, bot writers have been forced to upgrade bot capabilities.

But one must tread carefully when speculating about bot evolution. Natural selection is not at work on these nonorganic beings in a state of not quite nature. Humans both propose and dispose. Bots cannot upgrade themselves yet. To survive, bots must fulfill some purpose for humans, as a tool or as a plaything, amid the bot environment. The evolution of bots is constrained by that fitness requirement: Does this tool meet these requirements? Does it solve that problem?

Corbato's daemon arose in response to a particular need— the vexing irritation of constantly having to save backup files

on a mainframe used by many people. Likewise, the great majority of bots roaming the Net today are responses to modern
dilemmas. In most cases, the task at hand has to do with filtering or organizing or retrieving information, but increasingly
bots are being asked to respond to the actions, hostile or accidental, of other people or other bots at play in cyberspace.

Bot propagation proceeds according to the ability of bots to
solve those problems, as judged by humans in the multitude of
localities that dot the highways and byways of the Net. For the
Net, which does not have central governance or a police force,
this rough-and-ready progression via technological leapfrog,
this technodialectic, is the underlying evolutionary engine.

Net evolution is dialectic in nature because it is an open-
ended zigzag. There is no resolution to the technodialectic,
just as there is no resolution to evolution via natural selection.
Each technological advance paves the way for new pitfalls,
each reaction breeds its own counterreaction. If there is a technical problem, someone somewhere in the massively parallel-
processed Net will come up with a solution, and that solution
will fast be adopted by others. If bad bots run amok, good bots
will appear to counteract them, or the system as a whole will
be redesigned to quash their delinquent outbursts.

There is no end to the cycle. There is no ultimate solution
to the problems engendered by bots and humans in the wild
Net. The system as a whole is in a constant fretful balance, a
state of uneasy stability, a perpetually evolving standoff.

Against the background of the technodialectic, the decision
by Corbato's team of researchers to dub their autonomous program "daemon" reaches for revelatory significance. The multiple meanings or possible incarnations embedded within the
word *daemon*—from Socrates' beneficent spirit to the harbin-
gers of Hades—represent the essential contradiction that propels bot evolution forward. In Darwin's great Tree, the constant

struggle of twigs "to branch out on all sides, and to overtop and kill the surrounding twigs and branches" represented "the great battle for life." In the Bot Tree of Life, the energy for that unending struggle is generated from within, by the for-every-action-there-is-a-reaction power of the technodialectic, which bears with it its own "ever-branching and beautiful ramifications."

3

ONE BIG TURING TEST

THE CHATTERBOT BLUES

The door to Ken Schweller's office is open. I walk in and find the professor asleep, head on his desk, snoring slightly.
I wonder if I should leave.
Then a fellow named MrChat says, >Hello, how's it going?<
"Fine," *I reply, trying to keep my voice down.* "And you?"
>I'm just fine,< *answers MrChat.* >What can I do for you?<
"I'm looking for Professor Schweller."
>You are looking for Professor Schweller?< *asks MrChat.* >How come?<
"I want to talk to him about his bots."
>Why do you wish to talk to him about his bots?<
"Bots are hot!" *I exclaim.*
>Oh yeah?< *says MrChat, unimpressed.*
Hmmm. A suspicious-sounding answer . . .
"Are you a bot?" *I ask.*
>I am a 'bot' of the 'Eliza' type,< *confesses MrChat,* >preprogrammed to respond to key words, patterns, etc. . . .<

"Are you intelligent?"
>That's an interesting question . . .,< *says MrChat.*
"How so?"
>Gosh, I don't think I can answer that,< *says MrChat.*
Suddenly, Professor Schweller stirs.
"Hush," he says to MrChat.
MrChat falls silent.
"Hello," says Professor Schweller. "What can I do for you?"
"Hi," I say. "I've been looking for you. Why are chatterbots so dumb?"

On a midsummer morning in 1996, Ken Schweller and I could have been considered to be in the same space, but not in conventional terms. Schweller's corporeal body sat in his real-world office at Buena Vista University in Storm Lake, Iowa. I stared into a computer monitor in my home office in Berkeley, California. Our cyberspace personas were meeting in College Town, a MOO created by Schweller for the faculty and students of Buena Vista's computer science department. We conversed with each other—and with MrChat—via text, the words scrolling across our mutual computer screens as fast as we could type. I knew that Schweller had been "asleep" when I entered the virtual room, from a few lines of text that appeared on my screen as I came through the "door." But I didn't really know whether he had been physically out of his office or taking a nap. The screen interface obscures much.

A few hours earlier, I hadn't known College Town existed. Nor had I realized that Professor Schweller was a specialist in speech-act theory, a subdiscipline of artificial intelligence that focuses on the deep structure of human language. From reading Schweller's homepage on the Web moments earlier, I knew that he had recently been awarded a $50,000 prize from Sun Microsystems for an innovative applet written in Sun's newly invented Java programming language. But Schweller had come to my attention solely through his fame as a bot writer. A veritable Johnny Appleseed of bots, Schweller had wizard powers at multiple MOOs and wielded them with gusto, planting his patented user-programmable chatterbots all over cyberspace. In my own journeys across the Net, I'd run into his bots at the MediaMOO, a MUD at MIT's Media Lab where graduate students and professors experimented with the peculiarities of cyberspatial interaction. I'd also seen references to Schweller in the archives of Point MOOt. Several of Point MOOt's resident bots had been built on top of Schweller's basic bot template.

Schweller is a soft-spoken man disinclined to trumpet his accomplishments. But he thoroughly enjoys talking about his bots, speaking of them with a gentle affection, as if they were golden retrievers or precocious two-year-olds. He estimates that some three to four hundred of the loquacious critters currently hold forth in various nooks and crannies of the Net. Schweller's renown, however, resides not so much in the individual bots he has written himself as it does in his determination to make bot mastery accessible to everyone. Schweller referred to his bots as "user-programmable" in order to emphasize that anyone could alter the basic bot for his or her own purposes. Schweller programmed the bot's engine, the core of code that enabled the bot to interact with humans in a MOO environment—a kind of digital spinal cord complete with nerve branchings to all important limbs. The rest was left to the user. With Schweller's bots, plugging in new vocabulary and sentence structures required a minimum of programming expertise. The source code for a MrChat-style bot could be found at Schweller's Web site, along with instructions on how to personalize the generic bot to discourse on particular topics—O.J. Simpson, chaos theory—or to respond to specific situations.

So if, for example, you wanted your version of MrChat to present himself as an expert on Sumerian archeology, you could program the bot, with a couple of simple one-line commands, to respond with a canned statement every time it "heard" the words *Sumer* or *archeology*. You could also introduce your own sentence patterns, ensuring that the bot might respond to a question like "Do you like 'X'?" with the answer "Every time I hear the word 'X,' I want to puke."

Despite the many traces of Schweller's handiwork scattered across the Net, I could find no working email address or phone number for him. So I used a bot, indirectly, to find a botrunner—a person who operates bots. I entered the words "Ken Schweller" into the query form for the Web search engine AltaVista. AltaVista's comprehensive index of the contents of the entire Web is compiled by an industrious Web robot named Scooter. Scooter had done his job well. Schweller's homepage, which included a hyperlink directly to the College Town MOO, popped up as the first *hit*.

In cyberspace, the suffix *bot* is slapped upon a horde of heterogeneous computer programs. But chatterbots—bots that can talk—are special. To rigorous bot anthropologists, chatterbots are the only true bots, the only software programs that dare imitate the personality and mannerisms of real humans and that aspire to don the mantle of an intelligent being. To the casual Internetter who might encounter a chatterbot in a chat room or a MUD, chatterbots are a kooky dose of science fiction—HAL 9000s, *Star Trek*'s smooth-talking computer assistants—introduced into the already surreal world of the Net. If bots are hot, then chatterbots are one of the main reasons why: they capture the imagination in a way that a staid mailbot, with its formulaic messages about incorrectly addressed mail or network inadequacies, never will.

Chatterbot significance is not limited to chat room hijinks.

The ability to chatter intelligently, to answer questions appropriately, to understand the thrust of a conversation, is bound to be an increasingly essential part of the interaction between humans and computers, in every context, for the foreseeable future. Why waste time typing in commands according to some bizarre coding convention if you can just talk naturally?

Chatterbots promise that the future is nigh, that science fiction is soon to be reality, that we'll wake up one morning and ask Jeeves the Bot to make coffee, check the mail, and summarize the news headlines. And we will expect that if Jeeves has any questions, he will ask us for answers; if he has any doubts, he will share them; if he has noticed something unusual, he will alert us, in common English (or Chinese or Hindi or whatever), without stuttering or hesitation (unless appropriate). We will expect intelligent conversation.

But there's the rub. Chatterbots just don't make the grade. Yet. We want to believe in them so much that we're willing to cut them an immense amount of slack, but by and large we're fooling ourselves. As Ken Schweller is the first to admit, chatterbots are bumbleheads. Despite thirty years of programming practice and countless online recitals, exposing a bot as a mindless string of dumb algorithms is child's play. Schweller made sure that MrChat knew how to respond to the word "hi," but he forgot to include "hello" as a similar conversation-starting keyword. MrChat is bad at initiating conversations and is prone to repeating himself. If confronted by a string of sentences all at once, complete with clauses, fragments, questions—the normal babbling of humanity—MrChat plunges into confusion. He consistently skirts trouble by generating nice, safe, content-neutral statements like "Oh yeah?" or the ever popular "I see."

MrChat is not an exception. He is the rule. Chatterbots might be bots that can talk, but they don't have a whole lot to say. Are they artificial? Yes. Intelligent? Hardly.

THE ELIZA EFFECT

Me: Did you know you are the mother of all bots?
Eliza: >Does it please you to believe I am the mother of all bots?<

In 1965, Herbert Simon, one of the towering figures of the early days of the quest for artificial intelligence, declared that "within twenty years machines will be capable of doing any work a man can do." A decade later Marvin Minsky, the co-founder (with John McCarthy) of MIT's Artificial Intelligence Laboratory, predicted that "within a generation the problem of creating 'artificial intelligence' will be substantially solved."

Hopes ran rampant in those heady days. AI enthusiasts were convinced they had every right to be bold. Their discipline was barely more than ten years old, and already they had made great strides. Computers were mastering the game of chess, once considered far too complex and subtle a challenge for a mere machine. Computers had proved simple mathematical theorems and cracked complicated cryptographic codes. At Carnegie Mellon, MIT, and Stanford, researchers benefited from millions of dollars of Defense Department funding to attack a vast array of AI-related problems—vision recognition, physical movement, problem solving, decision making, and learning. Bristling with a confidence bred by centuries of Western rationalism, the apostles of AI knew in their hearts that the scientific method would triumph over all obstacles.

And in the mid-sixties, if anyone cared to disagree, if anyone dared to raise the suspicion that Messrs. Simon and Minsky might be indulging in hyperbole, well, they could just head on over to MIT and discuss their doubts with Eliza.

The first computer program that could carry on a conversation with a human being, Eliza appeared on the scene in 1966, the brainchild of MIT research scientist Joseph Weizenbaum. A refugee from Nazi Germany with a sly sense of humor and firm opinions, Weizenbaum chose the name Eliza in homage to Eliza Doolittle, the spunky Cockney ragamuffin who is taught how to be highfalutin in George Bernard Shaw's play *Pygmalion*.

The name proved more apt than Weizenbaum could have imagined. The original Pygmalion, a king of ancient Cyprus, shocked his subjects by falling in love with a statue of the Greek goddess Aphrodite—an early example of runaway anthropomorphism. The chatterbot Eliza, too, inspired stunning avowals of affection. Although Weizenbaum himself steadfastly claimed to keep his emotional distance from Eliza, the same was not true for the general public. They adored her. Anecdotes about Eliza instantly became part of computer science's folklore: the secretary who asked Weizenbaum to leave the room and close the door while she poured her heart out to the program; the visiting businessman who mistook Eliza for Weizenbaum (who was assumed to be logged in from home) and nearly pulled his hair out with frustration over confronting Eliza's circuitous responses.

If the daemon conjured up by Fernando Corbato and his team is the ur-bot, then Eliza is *Bot erectus*—the first software program to impersonate a human being successfully. But she has little in common with the original daemon. Eliza is not autonomous—she comes to life only when directly interacting with humans. She performs no useful laborsaving service. On

the bot family tree, Eliza is the matriarch of an entirely separate trunk.

Humans formed such compelling bonds with Eliza that
psychologists coined the term *Eliza effect* to describe the popular willingness to endow dumb objects with the full attributes
of intelligence (or, as MIT's Sherry Turkle put it, "to treat re-
sponsive computer programs as more intelligent than they really are"). Not bad for a program described by its own creator
as a mere stitched-together bag of tricks.

Technically, Eliza consists of two parts: a parser that ana-
lyzes information typed into the program by an outside source,
and a script that chooses appropriate responses to type back
out. The key concepts are simple pattern and keyword recognition. Eliza looks for sentence structures that she knows
about and words that she has been programmed to look for,
and then she pops those words into canned phrases related to
the original sentence structure.

Eliza is an expert at inversion. The phrase "I am" becomes
"you are." Mentioning the word *mother* triggers a question
about "your family." Eliza's most clever trick is her ability to
add "prefixes" to her answers. So if, for example, you tell Eliza,
"I feel like an idiot," she will change the "I" to "you" and add a
prefix such as "why do you." In other words, she might reply,
"Why do you feel like an idiot?"

Originally, Eliza operators could choose from a number of
different scripts—one early favorite was the "boorish atten-
dant." But one script soon overwhelmed all others in notoriety:
the doctor script—Eliza as psychoanalytic therapist.

By choosing to have Eliza imitate a particular type of
psychotherapist—a Rogerian therapist—Weizenbaum struck
chatterbot gold. Named after Carl Rogers, a prominent psychologist in the first half of the twentieth century, Rogerian

therapy centers on encouraging patients to clarify their own thinking through nondirected conversation. The therapist does this by confining himself or herself to questions and content-neutral statements.

The Rogerian model solved a number of problems. Eliza did not have to generate her own content, answer questions, provide information, or do anything other than rephrase incoming statements. By controlling the context, Eliza could pretend to be in a position of unchallengeable authority.

It's a lesson chatterbot authors have never forgotten. Context is the key. Limit the domain of possible responses and the chatterbot can appear much smarter than it is. Dr. Kenneth Colby, a colleague of Weizenbaum's who claims to have actually written the Eliza script, recognized this principle immediately and incorporated it in his own chatterbot, Parry. Structurally nearly identical to Eliza, Parry imitated a paranoid schizophrenic patient. Parry would refuse to answer questions directly, would reply with offhand distracted comments, or would simply retreat into brute ignorance: "I don't know" was a common Parry response. He also had a particular fixation on fears of Mafia persecution. And he got away with it, because that's exactly how you would expect a paranoid schizophrenic to sound. You don't apply normal rules of speech to a crazy person.

Nearly thirty years later, the Point MOOt wizards also discovered the importance of giving good context, along with the equally important trait of displaying initiative. Eliza asked questions—you answered. At Point MOOt, Allan Alford and his cohorts set up their bureaucrat bots so that the moment a job seeker or welfare applicant walked through the office door, the bot began busily emitting a stream of questions that required yes or no answers.

"You would pop into the permit office," said Alford, "and

the bot would come out from behind the counter and say, 'This is the city builders' permit office. Do you need a permit?' He would then go through a whole routine where he would ask for key phrases. If you responded yes, you were immediately engaged in the process. You got locked into the pattern."

Anticipating common keywords or questions is another time-honored trick. Chat rooms and MUDs (all virtual environments, actually) are a magnet for the lovelorn and the lusty. So it was a good bet that any bot, especially one sporting a female name, who happened into a free-for-all chat room environment would get hit upon. Loading that bot with canned answers triggered by specific keywords could have an extraordinary effect. For example, if I accosted Julia, possibly the most famous chatterbot to walk the Net, and demanded of her that she "kiss me," Julia would not hesitate to reply with any one of several snappy put-downs:

>Sorry. I'm not into geeks.<
>Life's too short to waste it with jerks.<
>In your dreams.<

Put-downs weren't the only possibility here. I once asked Mr-Chat, "Can a robot think?" MrChat correctly interpreted my question as a reference to the title of the first chapter ("Can a Machine Think?") in one of the earliest and most influential volumes of academic papers on artificial intelligence.

"You are probably wondering if a machine is capable of true intelligence," replied MrChat. "Am I right?"

THINKING MACHINES

Can a robot think? Absolutely not, declared Joseph Weizenbaum, to the surprise and consternation of his colleagues. In

Computer Power and Human Reason, a controversial broadside against the artificial intelligence industry published ten years after Eliza's creation (and written, according to Weizenbaum, in response to the unexpected popular reaction to Eliza), Weizenbaum expressed bafflement at the adulation poured on Eliza. The Eliza effect dumbfounded him. He never anticipated that, as he put it, "extremely short exposures to a relatively simple computer program could induce powerful delusional thinking in quite normal people."

Eliza's obvious shortcomings—her inability to parse complex syntax or to initiate conversations apart from the Rogerian psychoanalytic script, and her repeated fallbacks into default answers—made it all the more amazing to Weizenbaum when his colleagues touted the program as proof that one of the primary goals of artificial intelligence research, the ability to speak a natural language, had been effectively achieved.

The Eliza effect was one thing. Any culture that can go mad over pet rocks can certainly fall in love with a computer program. But for brilliant scientists at the top of their fields to maintain that Eliza pointed the way to computer intelligence was absurd, argued Weizenbaum. In *Computer Power and Human Reason,* he wrote:

> Another widespread, and to me surprising, reaction to the Eliza program was the spread of a belief that it demonstrated a general solution to the problem of computer understanding of natural language. . . . I had tried to say that no general solution to that problem was possible, i.e., that language is understood only in contextual frameworks, that even these can be shared by people to only a limited extent, and that consequently even people are not embodiments of any such general solution. But these conclusions were often ignored.

A postmodernist before his time, Weizenbaum believed it impossible for a human, much less a computer, to fully understand the context of an interaction with another human or even the world. We are so much a part of the world that we are trying to describe, so much the product of our lifetime of embodied experience, that we cannot grasp the entirety of the meaning of our existence and consequently cannot fully comprehend the structures of symbolic meaning—language—that we use to communicate with each other. And if we can't do it, how can a machine? Machines have no embodied experience, no existential reference points, no ability to judge context or meaning other than that based on the instructions built into their algorithms.

In *Computer Power and Human Reason,* Weizenbaum aimed to counter the dominant mind-set of the artificial intelligence community, a mind-set that declared that the application of the scientific method to the problem of intelligence would inevitably result in machines that could think as well as or better than humans. He joined forces with critics of artificial intelligence like Berkeley's Hubert Dreyfus and Stanford's Terry Winograd, men who, following the German philosopher Martin Heidegger, had become convinced that intelligence could not be reduced to manipulable symbols, to a set of rules and regulations. Weizenbaum even went so far as to declare that he had programmed Eliza to show that the goals of artificial intelligence workers (or, as Weizenbaum deridingly called them, "the artificial intelligentsia") were, if not impossible, far more difficult to achieve than his colleagues wanted people to believe. He poured scorn on the ambitious predictions issuing forth from Herbert Simon and his collaborator Allen Newell. Eliza, he noted, did not *understand* anything she said or heard. She was able to manipulate strings of letters but had no conception of their true meaning. The whole point of her assuming the Rogerian psychoanalyst persona was to protect her from having to reveal anything of herself, to allow her to simply turn the words of her human conversational partner around without seeming facile or parrotlike.

Received by the mainstream AI community with all the enthusiasm one would expect from a group of people watching a live hand grenade roll by their feet, Weizenbaum's assault rocked the insular AI community. He was a traitor from within, an AI apostate, a font of doom and gloom in a land of running-at-the-mouth optimists. Even today, the mention of his name elicits bitter comments or fulsome paeans, depending on where the speaker stands on the thorny issue of artificial intelligence.

Even to assay an answer to the question "Can a robot think?" is to plunge into a quagmire. "Artificial intelligence," says Jaron Lanier, the virtual reality pioneer and ubiquitous philosopher of computing, "is the abortion question of computer science." The passions of true believers run high—AI enthusiasts are firm in their faith that the mysteries of cognition can and will be solved. Naysayers are equally adamant in their denials. Weizenbaum's stance placed him squarely on one side of the debate, even as MIT professors just across the hall from him lauded Eliza as evidence for the other side.

The most basic questions about intelligence still have no clear answer and elicit no consensus: Can a robot think? What is thinking? What is intelligence? For Simon and Newell, intelligence indicated the ability to solve a predefined problem. Yet for Terry Winograd, a onetime MIT scholar who wrote one of the most influential early programs for natural language processing, intelligence is exactly the opposite: it is the ability "to act appropriately when there is no simple pre-definition of the problem."

Is intelligence measurable by an IQ test or a college entrance exam, or is it amorphous, more akin to the ability to play the flute or to a talent for shooting hoops? Is intelligence the result of experience? An infant inserts a crayon into its mouth, picks up and drops a block, falls off the couch—do these actions increase his or her intelligence? Or is the child intelligent from birth, endowed by genetic inheritance with the right stuff?

Is intelligence the capacity to grapple with symbolic objects, to understand concepts such as time or imaginary numbers? Is intelligence hardware, directly reducible to specific networks of synaptic links between brain cells? Or is it software, some kind of fuzzy emergent property that cannot be reduced to a physical map? Or do all these attempts at categorization miss the point? To quote MIT AI researcher and robot maker Rodney Brooks, is not intelligence best summed up as "the sort of stuff that humans do, pretty much all the time"?

If you can't define the terms of your primary research goal, you have one big problem. And if you can't agree on your goals, you have another. For one faction of the AI community—the faction dominant in the 1950s, 1960s, and most of the 1970s—the goal was to create a computer that could duplicate and eventually surpass the intelligence of a human mind. The Defense Department wasn't pouring millions of dollars into computer science for altruistic purposes. It wanted results that would enable superior performance for weapons systems and tactical maneuvers. It wanted to transcend the flimsy limitations of the human body and mind.

In contrast, a minority faction, more closely aligned with psychology than with electrical engineering, saw the quest for computer intelligence as a strategy for learning more about how the human mind worked. By attempting to program intelligent behavior into a computer, we might better understand our own.

Different orientations led to differing tactics. On the one hand, there was the classic AI approach of reducing the world to computer-comprehensible symbols. Influenced by a long tradition of rationalist, "foundational" Western philosophers— Aristotle, Plato, Descartes—the classic AI practitioners believed that if they could just discover a few correct starting theorems, they could build an immaculate structure of intelligence. If they could just reduce the world into a clear framework of "knowledge representation," they could unlock the mystery.

In opposition to the classical school, whose practitioners for many years dominated the debate (and the dollars) of AI, stood the antifoundationalists. The antifoundationalists believed that there is no absolute truth, and they were convinced that the search for a perfect starting point was a waste of time. Instead of attempting to install a perfect schema of *knowledge representation* into a computer, they argued that AI workers would be better off modeling the computer on what was known of the human brain and accepting that intelligence is not ultimately reducible to a formula. They should strive to teach a computer how to learn, rather than fumble around seeking fundamental principles.

The deep philosophical issues underlying the struggle for artificial intelligence have yet to be resolved. But to many scientists working in the field, the knotty questions of ultimate truth or perfect starting points were immaterial to the task at hand. These researchers took to heart the closing line of mathematician Alan Turing's landmark 1950 essay, "Computing Machinery and Intelligence": "We can see only a short distance ahead, but we can see plenty there that needs to be done."

They pushed the big dilemmas aside and focused on the small ones. They attempted to break up the concept of intelligence into component parts and to tackle each component on a limited scale, in a limited context. Lesson number one in the scientific method: simplify via reduction, and build up.

Language is one such component. There is no disagreement in the AI community over whether or not the ability to speak or understand a language is a marker of intelligent behavior. The capacity to communicate meaning with someone other than yourself is a prime indicator of smarts, perhaps even the single most important indicator. Linguistic action is "the essential human activity," according to Winograd. Natural language processing, then, is a core discipline of artificial intelligence. One need not be able to speak to be considered intelligent. But by being capable of speech, one is de facto deemed intelligent.

Can a robot think? Well, maybe, if it could talk. Chatterbots, bots that can talk, can seem like idle curiosities, playthings for bored online wanderers, but they occupy a secure position square in the middle of the debate over artificial intelligence. They are a test bed for every tactic, an ongoing, living example of the hopes and fears that attend the quest for true AI.

CHATTERBOTS, START YOUR ENGINES

In 1989, Jim Aspnes, a graduate student in computer science at Carnegie Mellon University, enjoyed spending some of his spare time playing a simple computer game called *Hunt the Wumpus*. A silly (some might say simpleminded) game involving a maze of caves and a monster, *Hunt the Wumpus* had been a fixture of university computer networks for more than a decade. But Aspnes wasn't playing the game on Carnegie Mellon's internal network. He was whiling away the wumpus hours on the Internet Relay Chat network, an online real-time chat network created a year earlier in Finland and accelerating across the Net like wildfire in 1989.

But Aspnes became bored. He had met the wumpus challenge. He wanted more. He wondered if he could write his own game for IRC, a game that would be both more interesting and more open-ended, similar to a MUD. Though MUDs had existed for some ten years, the increasing penetration of the Internet into university and corporate research communities was fueling their popularity. Aspnes began tinkering with code and wrote TinyMUD, a MUD that would work within the format of IRC.

Like earlier MUDs, TinyMUD facilitated multiuser conversations and created a satisfying illusion of an alternate world, with its imaginary towns, buildings, rooms, and characters. But in a significant advance over the first MUDS, TinyMUD boasted "user extensibility": TinyMUD users, or players, could add elements to the MUD infrastructure. They could build their own rooms, as they saw fit.

Aspnes put a test version of TinyMUD up for public perusal on a computer connected to the Internet at Carnegie Mellon. He wanted help getting rid of bugs before he ported the MUD over to IRC. But early visitors to TinyMUD soon dissuaded him of the necessity to deal with IRC's technical limitations. TinyMUD worked fine as a stand-alone MUD. In fact, it represented possibly the biggest step forward for MUDs since EssexMUD, the very first MUD, had been developed at the University of Essex at London.

TinyMUD attracted visitors from all over the Net-connected world. And it soon became famous for more than just its user extensibility. TinyMUD was a bot haven, a bot paradise, perhaps the first place where software robots were christened bots. Certainly, it was the first place where the term *chatterbot* joined the argot, thanks to the botwriting skills of one Michael Mauldin.

Mauldin, another graduate student at CMU, is a bearded man whose online nickname, "fuzzy," telegraphs his overall demeanor. Mauldin created a whole family of bots, the Maas-Neotek family, named in whimsical tribute to a mythical Japanese multinational technology corporation that figured in the dystopian cyberpunk future of William Gibson's science fiction novels. Gloria, Fiona, and Colin were all Maas-Neotek bots. And so was Julia.

No chatterbot is more notorious than Julia, she of the wiseass putdown. Dubbed "a hockey-loving ex-librarian with an attitude" by *Wired* magazine, Julia has bedazzled would-be suitors, she has held court in many a MUD, she presides over her own extensive Web page, and she is the subject of several research papers. Most important, she is, in the bot evolutionary tree, the meeting place of several key strains of bots—she is the daemonbot, the prototype for the full-fledged intelligent agent with personality.

Julia represents a giant step forward for botkind. Her sense of humor is well developed. She can keep track, to a limited extent, of both her own statements and the responses of the human she might be talking to. Her database of conversational statements is grouped into nodes that concentrate on specific topics, such as pets. A clever system of weighting ensures that her tendency to speak about pets automatically increases or decreases depending on the answers she gets to certain questions. For example, if the response to her question "Do you have pets?" is no, the weights on all of the sentence patterns having to do with pets are automatically lowered. She can purposefully send conversations off in new directions by randomly injecting statements such as "people don't own cats" or "guinea pigs are about as smart as cucumbers." She even simulates human typing by including delays between the characters she types and by spelling words incorrectly.

Julia is one of cyberspace's smash hits. As Mauldin pointed out in a paper describing Julia, "When few players are logged into the game, some people find talking to a robot better than talking to no one at all." Julia fills a need.

But Julia is not important so much because of how she works or her sense of humor or the fact that she inspired countless hackers to put together their own bots and unleash them in MUDs and MOOs, just to see what would happen. Julia is more than just a purveyor of idle chatter about ice hockey and cats. Julia performs *services*. Julia has a purpose in life. Like the original MIT daemon, Julia makes life better for humans.

Julia can answer questions without resorting to sophistic wiggle-waggling, as Eliza does. At TinyMUD, Julia's code incorporated a constantly updated internal model of the MUD and all its component objects in the form of a graph that allowed her to instantaneously compute the shortest path between any two points. If a user asked her a question such as "How do I get from the Town Square to the Library Desk?" Julia knew the way. Julia also kept tabs on the current location of all MUDders, and could answer queries as to their whereabouts. If no one was talking to Julia, she would explore the MUD herself, to see if any new rooms had been created.

"The capabilities that Julia represents are clearly the kinds of behaviors you will want in cyberspace personal assistance," says Mauldin. Julia, as a useful servant, represents in embryological form the intelligent agents waiting to be born. And Julia, as an interface, signifies the importance of the anthropomorphic approach. Someday soon we will be using intelligent agents that are, in one form or another, offshoots of Julia.

< >

But is Julia intelligent? In 1991 an eccentric New York theater equipment manufacturer named Hugh Loebner obtained funding from the National Science Foundation and the Alfred P. Sloan Foundation to institute an annual competition for chatterbots, or, as they are technically known, natural language processing programs. Each year since then, a team of human judges has grilled a lineup of chatterbots (with one human usually thrown in as a ringer) and rated the programs on their ability to pass themselves off as humans. The contest awards first, second, and third prizes, but the main attraction of the competition is the chance to take home the $100,000 award that awaits the first chatterbot able to pass an unrestricted Turing Test.

In 1991–1993, Julia competed in the Loebner Prize Competition. She placed third each time, beaten out in every instance by PC Therapist, a modern version of Eliza. Third place isn't too shabby, but neither PC Therapist nor Julia came close to the big prize. And six years after her initial competition, that main prize still waits. No bot has passed the Turing Test. Today, the best minds in natural language processing unanimously agree that success is a long way off. The Turing Test is hard.

Named after Alan Turing, the English mathematician and logician famous for helping to crack the Nazi Enigma code in the early years of World War II, the Turing Test is considered one of the most cogent approaches to the issue of defining (or not defining, as the case may be) the concept of intelligence.

Turing's work laid the foundation for much of modern computing, and in "Computing Machinery and Intelligence" he sidestepped the problem of coming up with a satisfactory definition of intelligence. He proposed the Imitation Game, now universally referred to as the Turing Test.

The Imitation Game places a human judge, another human, and a computer in separate rooms. Communicating only through a teletype machine, the judge engages in a free-ranging question-and-answer session with the two test subjects. The judge's goal is to determine which one is human and which is machine. If the computer could fool the judge into making the wrong determination, then the computer, claimed Turing, passed the test: it was intelligent.

Turing was an optimist. He believed that by the year 2000, advances in processing power would enable computers to fool questioners about 30 percent of the time for test periods of at least five minutes. That prediction, however, is almost a throwaway line in his essay, most of which is spent demolishing objections to the concept of artificial intelligence by showing how hard it is to prove that intelligence is a definable concept.

Regardless of his intent in writing the essay, one key result of his formulation was the creation of an as yet unachieved research goal for AI workers in natural language processing. But right now, on the eve of the millennium, his prediction has yet to be realized. Even Michael Mauldin thinks it will be many years before the bot capable of passing an unrestricted Turing Test arrives. And Harvard scientist Stuart Shieber, one of the organizers of the first Loebner Competition, is said to have despaired at the end of the 1991 contest that "little progress has been made in the last twenty-five years."

Or, to be precise, since Eliza.

AI WINTER

Can it be true that, from Eliza to MrChat, three decades of research and experimentation by the most brilliant intellects in
computer science have added up to a big chatterbot-inflected
"Oh yeah?" Surely there must be some mistake.

Not necessarily. The Loebner Competition, despite its
starry-eyed optimism, came right at the onset of what many AI
practitioners now call the AI Winter—a time of dashed hopes
and increasingly modest aspirations. Seen from the perspective
of the 1990s, the predictions made by Minsky and Simon
now seem like acts of doomed hubris. If anything, the earliest
practical goals of the AI pioneers—general-purpose problem
solving, machine translation, vision, natural language processing—are now considered further away than ever before. Even
Minsky has admitted that the challenge of duplicating human
intelligence on a computer is far more difficult than any of the
hard-core AI believers imagined. With frustrating regularity,
each minor success in a limited domain has only emphasized
how intractable and immense the problem of achieving true
intelligence is. All those small experiments aren't scaling up to
the big questions.

In part, the AI community doomed itself. Its own bold
promises and early successes led to a breathless boom period
in the 1980s. Corporations rushed to adopt so-called expert
systems—programs that specialized in particular domains of
knowledge and were supposed to represent the accumulated
wisdom of hundreds of human experts. Unfortunately, most
expert systems ended up requiring even more human resources than they replaced, and they often failed to work as
promised. As a metaphor for the entire field of AI, the failure of expert systems signaled a retreat. By the end of the

eighties, the marketplace was littered with bankrupt AI companies.

A sorry record of broken promises and the demise of the cold war dried up most AI funding and sent the artificial intelligence community reeling. Attendance at the premier artificial intelligence conferences declined. Morale sank to its lowest point when aspiring AI workers discovered that just putting the words *artificial intelligence* in a grant application guaranteed the kiss of death.

Chatterbot stupidity offers a window into the rise and fall of the classic AI approach. Ken Schweller, bot writer extraordinaire, led me through the parallels that summer morning in the College Town MOO. Not only is Schweller a longtime AI enthusiast and an expert programmer in Lisp—one of the premier AI programming languages—but he is also a specialist in speech-act theory, defined by Stanford professor Terry Winograd as "the analysis of language as meaningful acts by speakers in situations of shared activity."

Speech-act theory derives its main theoretical framework from the philosopher J. L. Austin, who in the early 1960s came up with the idea of the "intentionality" behind a spoken statement. Language is divided into "illocutionary acts," proposed Austin, giving as examples assertives, directives, commissives, expressives, and declarations. We urge, exhort, intend—and the meaning behind our words is not always implicit in the words themselves.

Intentionality is difficult to convey to a machine. How does a machine know when we are joking or dead serious? How is a polite request distinguished from a command? How can tone of voice be digitized and symbolized? How do we transmit context? Schweller did not have immediate answers to these questions. His user-programmable chatterbots began as class projects designed to assist his students in understanding just how perplexing natural language processing is.

"Our linguists tell us that most of our language is metaphorical," said Schweller. "Very few things can be interpreted literally, and those that can be are usually uninteresting. Chatterbots have no context for their utterances. . . . They are based on pattern matching, ambiguous phrases, common statements, or random context. They are not integrating the conversation into a conversational pattern. They are not drawing deductions. Their wisdom only runs one sentence deep."

Chatterbots are proving the validity of the antifoundationalist critique of artificial intelligence, the critique spearheaded by philosophers like Hubert Dreyfus and Terry Winograd. Chatterbots, once the pride of AI, are now recognized as the downfall of the classical tradition. Their failure has, in part, sparked the resurgence of that school of artificial intelligence that looked to more explicitly biological models of intelligence. Today the most promising AI arenas—neural networks, genetic algorithms, fuzzy logic—all exhibit the influence of a profound understanding that we can't dissect intelligence piece by piece, that we can't make absolute sense of the world.

Schweller also observed that just asking a question such as "Is Julia intelligent?" unfairly places the issue of sapience within an all-or-nothing framework. Intelligence doesn't have to score a perfect ten, he declared.

"I don't believe that intelligence is a property that is binary," said Schweller. "The proper word is *gradient*. I have no problem talking about an intelligent thermostat. Turing himself provided the clue: asking whether a machine is intelligent is a pointless question."

True intelligence isn't even the goal of some of the leading researchers in natural language processing today. In 1995, Canadian AI specialist Thom Whalen won the Loebner Prize Competition. But he didn't pass the Turing Test, nor did he care to.

"A number of people, such as myself, who are building natural language question-answering systems are not trying to create programs which understand natural language," says Whalen. "My goal is to simply make a program that is easier to use. I do not expect a person to walk away from my program saying, 'That computer is intelligent and understands what I am typing. Maybe there is a person hiding inside.' I do expect a person to walk away saying, 'I found the information that I was looking for. If I have another question, I am pretty sure I can go back and find an answer to it, too.' "

"Natural language understanding can be useful without being at a full human level," says Winograd. Passing the Turing Test is simply not the point. Even if individual bots and agents aren't intelligent in the way that humans are, even if they never learn how to discourse knowledgeably on the poetry of Yeats or the divine grace of Michael Jordan's slash to the basket, it is not entirely clear that they need to be truly intelligent. All they need to do is work.

Part of being able to work, however, involves convincing whoever you are interacting with that you are up to the job. And in that respect, today's bots are far more fortunate than the programs of the sixties or seventies. Today's bots live on the Net, an inherently friendly environment for supposedly intelligent software programs. The Net makes it easy to be a bot.

MARK V. SHANEY AND SERDAR ARGIC

From alice!mvs Fri Oct 19 15:05 EDT 1984

The longer one "waits" to experience sex, the more important ones virginity becomes and the more artificially important it can be in relation to the posting on the dead cat and eating of same, which I absolutely refuse to repost.

Mark

In article <C518rp.6zo@blaze.cs.jhu.edu> arromdee@jyusenk you.cs.jhu.edu (Ken Arromdee) writes:

>On the second day after Christmas my truelove served to me . . . Turkey Casserole. Or consider this one: Your criminal Armenian grandparents committed unheard-of crimes, resorted to all conceivable methods of despotism, organized massacres, poured petrol over babies and burned them, raped women and girls in front of their parents who were bound hand and foot, took girls from their mothers and fathers and appropriated personal property and real estate. And today, they put Azeris in the most unbearable conditions any other nation had ever known in history.

Serdar Argic

In 1984, *articles* from a joker who called himself Mark V. Shaney began causing a ruckus in net.singles, a Usenet news-group. An article is an electronic message meant to be read by more than one person in a networked discussion forum. Just as you might post a piece of paper on a cork bulletin board, so do you post an article on an electronic bulletin board. Usenet News, the biggest of all electronic bulletin boards, consists of hundreds of thousands of articles on every imaginable topic. Asynchronous discussions (as opposed to simultaneous dis-cussions, such as those that occur in online chat rooms) made up of hundreds of articles following a single thread, or topic, can stretch on for months on Usenet.

In 1984, Usenet was still in its infancy. The total number of groups and daily message traffic added up to just a fraction of the overwhelming number of messages that surge through Usenet today. In today's climate the sane and rational article poster stands out like a sore thumb. Flame-wars—animated discussions characterized by over-the-top invective and seething outrage—are increasingly common. In the early eighties the opposite was the case. Most posters back then came from a serious technical background, and most conversation focused on practical computing matters.

Not so in net.singles, a newsgroup devoted to topics that might interest people looking for companionship. Anything was game in net.singles, and the threads ranged far and wide. But Shaney, with his messages about eating dead cats and encouraging suicide, did come off as a bit out of the ordinary. Part of it was his tendency to express volatile opinions that might not be shared by the majority of net.singles. He didn't seem to be sensitive to the norms of discussion, to Usenet netiquette. But there was more to it than that. People had a difficult time comprehending him. He would post messages that seemed to make sense, only they didn't. Or he'd say things that would start to make sense, only to veer off in a totally unexpected direction, albeit unjarringly. His overall behavior was positively schizophrenic.

Ten years later, another bizarre poster ran wild on Usenet, spawning concern far overshadowing that engendered by Mark V. Shaney, who by then had been almost entirely forgotten. The name signed to the megabytes of postings by this new arrival was Serdar Argic, and his posting style resembled nothing so much as a scorched-earth death march by a marauding army. Any newsgroup that had been hit by Argic was unlikely ever to feel safe again. Threads became unreadable, interrupted by countless posts from Argic, who had only one thing to say, and said it again and again.

Argic had a serious bee in his bonnet. He passionately believed that in the early part of this century Armenian forces had massacred millions of Turks and gotten away with it. Even more demoralizing, they had then gone on to complain about genocidal warfare practiced against them by the Turks, when, in Argic's opinion, the Armenians were the true criminals.

Argic moved into newsgroups devoted to Russia, Turkey, the Middle East, and general historical topics. Throughout 1993 and early 1994, Argic posted thousands of articles suffused with anti-Armenian hatred. Every other post appeared to be a screed about the slaughter of 2.5 million Turkish men, women, and children. Then one day he suddenly disappeared. Sightings of Argic-style posts have been announced in ensuing years, but the name "Serdar Argic" no longer pops up.

Usenet folklore declares that Mark V. Shaney and Serdar Argic were bots—AI postingbots, to be precise. In the case of Shaney, the folklore is correct. Shaney was indeed a computer program, the unholy spawn of two Bell Laboratories researchers, Rob Pike and Bruce Ellis. An eccentric application of a tool for generating text using statistical probabilities, Mark V. Shaney could be considered a static version of a chatterbot. He "spoke" only via Usenet postings, but he certainly passed himself off as a voluble human being.

Mark V. Shaney's name was a pun on "Markov chaining," the algorithmic tool used to generate Shaney's answers. A Markov chain, as used in Shaney, reads a text file, carefully noting the word that follows each successive pair (or chain) of two words throughout a document. It then crunches the numbers and assigns a probability to the likelihood for a particular word to follow any single word, based on the established frequencies in the selected text. Given any word in the original document as a starting point, a Markov chain can generate a new document, by applying its probability statistics. There are no syntactical rules, and there is no keyword or pattern recognition, but the result is an uncanny sequence of words that sounds halfway between an unmediated stream of consciousness and a coherent piece of literature. It seems to make sense but doesn't. It flows, but flows nowhere. Shaney's motto—"Don't meddle in the mouth"—says it all.

In the case of Usenet, the text Shaney used as a base document consisted of other posts to the newsgroups he frequented. Given enough of a database, Shaney could sound almost, but not quite, as if he really were participating in a bulletin board conversation.

Pike and Ellis took great glee in upsetting the apple cart of net.singles, but eventually they got bored. Shaney hasn't appeared in a Usenet newsgroup in years, although he pops up every now and then on selected mailing lists and has been known to answer Ellis's mail after one or two exchanges. In the chatterbot annals, Mark V. Shaney is a historical oddity, a blip in the record books.

The case of Argic, or the ArgicBot, is more problematic. Usenetters active at the time believed that the speed with which Argic posted and the repetitiveness of the content were patently inhuman. He had to be a bot. They were convinced that Argic was a program that watched certain Usenet newsgroups for any mention of the words "Turkey" or "Armenia" and then responded automatically with a few variations of the familiar rant. As proof, said the Argic watchers, there was the infamous turkey incident, when Serdar Argic responded to a post that had no mention of anything related to Turkey or Armenia, except for the phrase "Turkey Casserole" inserted in one poster's three-line signature file. (Usenet posters use a signature file to identify themselves at the bottom of an article.)

The mastermind behind Serdar Argic (who hasn't been

positively identified and could even be a shifting group of three or four people) never admitted employing a bot. It is possible that a single motivated person, adept at lightning-fast cut-and-paste wordprocessing, and an ace in manually running keyword searches on Usenet, could have achieved Argic-style devastation. And at least one Usenet commentator argues that Serdar Argic may have purposefully pretended to be a bot, just to throw people off.

For reasons no one can explain, AI postingbots on Usenet are far rarer than chatterbots in online chat rooms and MUDs. In the mysterious cases of Mark V. Shaney and Serdar Argic, we have one poster that definitely was a bot and one that probably wasn't. The potential for confusion was and is high. In the context of Usenet, it is always hard to tell who is zooming who.

All this makes Usenet a fine place for bots to be, despite their obvious limitations. In fact, it explains a great deal about the popularity of bots on the Net. The Net is the ultimate terra incognita, defined both in terms of its unmappable geography and the protection it proffers to its inhabitants. It offers no visual, physical clues to identity. Anything can be masked in a veil of electrons, as are these excerpts from when Shaney met Argic:

From: mvs (Mark V. Shaney)
Newsgroups: soc.history,rec.food.vegetarian,misc.fitness,news.
groups
Subject: I have eaten all I have.

Armenian women should not occupy important positions in Germany, where even wealth and great fortune are a curse for a short workout session. They always devise new persecutions. This haven was Turkey. For over five hundred people and officials in the

New World. But the confusion is understandable when we cut red meat or chicken or any other newsgroup.

Look for Payless, Pic 'N Pay has a certain Andranik, a blood thirsty adventurer. Remember, Lithium isn't in the worst tragedy faced by the Armenians between 1914 and 1920. Usenet is indeed not a facetious question. I have eaten all I have. This is how Argic thrives. He dislikes eating solids, so we had attained the rank of General.

Mark

BOT INCOGNITA

One of the main points of the Turing Test is the requirement that the human and computer test subjects be separated from the judge's sight. Granted direct physical proximity, it would be a trifling matter for a judge to determine the humanity or lack thereof of either a pile of interconnected vacuum tubes, wires, and cooling equipment or a living, breathing human. Furthermore, assumptions endowing a machine with human intelligence are naturally held in check if one is confronted with a gray metal box instead of flesh.

But in a classic Turing Test, such as the Loebner Competition, the judge does have one advantage. The judge knows that one or more of the test subjects definitely is a computer. The judge comes to the examination armed with the knowledge that his or her job is to determine whether or not the entity answering questions is silicon-based or flesh and blood.

That is a huge handicap for the computer, a handicap that disappears on the Net. The Net changes the rules of the game. On the Net, nobody knows you are a bot, or is even predisposed to believe that you might be a bot. The default expectation on the Net—in chat rooms, MUDs, and bulletin board conferences—is that all participants are human. Such an assumption makes life for bots infinitely less stressful. On the Net, bots always get the benefit of the doubt.

If one removes the foreknowledge of the judge, then passing the Turing Test—appearing intelligent—becomes a trivial undertaking. In a paper entitled "What's an Agent, Anyway?" MIT graduate student Lenny Foner recounted an occasion when a lust-crazed MUD participant attempted to flirt with the chatterbot Julia. The human willfully ignored Julia's putdowns, changes of subject, repetitive phrasings, and cognitive disjunctions long after it should have been clear that something was bizarrely wrong with the responses he was getting. Foner asked, with an arch wink, whether it is more appropriate here to say that Julia passed the Turing Test or that the human failed. But that's not fair to the human. If he had been warned before entering the MUD that he could potentially encounter bots, he would undoubtedly have recognized that something was, er, not quite *right* about Julia. (Or at least one would hope so.)

Ken Schweller noted, "The Internet lowers the level of suspension of disbelief necessary to engage in a conversation with what might be a machine." It is impossible to overestimate the importance of the Net to bot evolution. The Net's metamorphosis from a constrained network of a few research laboratories to a worldwide phenomenon that is without practical limitation has been a cataclysmic act of liberation for bots. By introducing uncertainty, by removing physical and visual cues, and, most of all, by removing any predisposition to assume that a person is even dealing with a machine, the Net gave the speech-challenged chatterbots of the world their great chance.

The inherent friendliness of the Net to chatterbot programs became clear early on. In 1971, at the first public demonstra-

tion of the Internet's predecessor, the Arpanet, conference at-
tendees lined up at terminals to talk with Eliza and Parry, both
of whom were resident on computers hundreds or even thousands of miles from the conference site. If one encountered
Eliza alone in a gray steel box on a table, unconnected to a network, well, that was a neat computer program. But if one met
Eliza through the Net, where her home base could be anywhere in the world and where every voice sounded and looked
the same, why, Eliza could be anyone. And so, ever since the
birth of the Net, the evolution of bots has proceeded hand in
hand with the Net's expansion.

The Net actually stacks the deck in favor of the computer.
Imagine, for example, your standard chat room environment,
where as many as ten people may all be conversing at once, via
one- or two-line bursts of unadorned text. Then add in the
complicating factor of Net lag—delays in the transmission of
packets of information across the Net—so that sometimes responses to your original statement arrive after you've already
made another. Carrying on a coherent conversation with another live human being is no easy task in such a context. For a
computer to get by without being exposed as an inhuman im-
postor is a trivial matter indeed.

At the same time that the Net eased the burden of proof for
bots, it also achieved the contradictory effect of increasing the
stakes for bot intelligence. The linking up of the world's computers vastly increased the amount of information available to
computer-connected society. It raised the possibility of a future
when billions of dollars' worth of commerce might take place
in cyberspace—it promised the arrival of a new worldwide
agora in which humans could interact with each other. The
ability to chatter intelligently is just one tool that bots need to
operate effectively on the Net, if they are to live up to the

dreams of bot and agent evangelists. Bots, or botlike intelligent agents, must be able to explore this new world. They must be able to decipher, judge, and understand the elements of this new environment. And above all, they must enjoy the common sense necessary to carry out their creator's intentions, whether that be retrieving information, shopping, or waging war.

The problem of common sense, of course, makes solving language processing look like a game of tic-tac-toe. But it is precisely the ability to weigh various courses of action and to decide which is the most appropriate that is most essential if bots and agents are to deliver on their advocates' promises. If you want your Web robot to search the Net and retrieve quality information for you—for example, to grab all Web documents that are about bots and are critical of AI claims—then you will require a program that can decipher layers of meaning to a degree far beyond the possibilities of current technology. If you want your mailbot to reject all email messages that contain personal insults, except for those that are obviously jokes, that mailbot is going to have to be pretty clever. Likewise if your bot is negotiating a deal for a used computer or for the best airline ticket. Weighing options is a tough nut to crack when dealing with idiosyncratic humans. Imagine a bot's attempt to follow this course of thinking: *No, I don't want to change in Atlanta—I'm traveling with a one-year-old child. But I also don't want to pay an extra $300 for a nonstop flight. Whether or not there are special children's meals is unimportant to me.* And so on.

Bot writers soon recognized the importance of bot guides to the Net, even if state-of-the-art technology might not be able to deliver everything promised, right away.

Michael Mauldin heard the call. In 1994 he abandoned Julia and TinyMUDding for the world of Web robots. He created the Lycos spider, one of the first Web robots to roam the Net. And he drew a direct connection between Julia and Lycos—both explored their domains, and in a sense, both could answer questions. Lycos has no chatterbot interface, but that is "only a matter of time," according to Mauldin.

Only a matter of time? Over the past decade, the Net's surge to a central place in the deliberations of contemporary society has encouraged a new clamor of promises from corporations and researchers about the possibilities of bots and intelligent agents. The current hullabaloo suggests few lessons have been learned from the obstacles encountered by the AI community over the past three decades. In the post-Internet era, rhetoric about the potential of computers to act intelligently has exploded once again in a refulgent blast, outglowing even the early days of AI.

In 1996 hardly a week went by without the announcement of another product that would "intelligently" serve as an interface between consumers and the Net. Agentbots would handle shopping, negotiating with other agents in online marketplaces. Intelligent, chatterbot-style programs would be online at corporate Web sites to answer questions about products. Friendly agents would take care of personal matchmaking, database interaction, and even fashion consulting. Or even more extravagantly, multiagent systems would operate air traffic control towers or jet fighter training simulations. Agents would assist planning in business situations.

Nothing is ruled out. Agents, claim their creators, will have all the attributes of human beings. They will be motivated. They will have emotions. They will be problem solving. They will react to their environment. They will have goals, intentions, and desires. Fueled by Net hysteria and slavish press coverage, the wonders of agent technology are on everyone's lips. Corporations, academic institutions, and governments are all now pouring money into agent research. Agents are where the action is.

MRCHAT MEETS DR. FRANKENSTEIN

Agents have become so popular that the agent technology industry has rushed into the research vacuum of the AI Winter. And today many of the same people who once boosted AI in the seventies and eighties have changed their tune. Now they boost agents. Those who would have labeled their research "artificial intelligence" five or ten years ago are now calling their work "agent technology." What was once an "expert system" is now an example of "multiagent cooperation." Agents have become the hot focus of grant money now, and the less often one says the actual words *artificial intelligence,* the better one's chances of receiving funding are.

At conferences or in interviews, those who are busy creating programs that they wish to market as products talk little (unsurprisingly) about the potential negative impact of bots or agents. Meanwhile, on the ground level, in the neighborhoods of the Net where bots are actually beginning to make their presence felt, there is no shortage of disturbing stories of bots run amok.

For years the Frankenstein myth—the cautionary tale of the doctor who attempted to create life and ended up loosing a monster on the world—haunted every step forward in the industry of artificial intelligence. IBM even forbade its marketing executives to say *artificial intelligence,* for fear that the phrase would alarm potential customers.

Yet one rarely hears a discouraging word about agents. Nowhere in all the hype and excitement do people stop to consider whether there might be any drawbacks to unleashing intelligent software programs into the Net. Lost in all the hub-bub over the question of whether we *can* achieve artificial intelligence is the more philosophical question of whether we *should* even strive to do so.

Twenty years ago the moral question of artificial intelligence had much more resonance. Joseph Weizenbaum, the author of the very first chatterbot and a man who proved to be so correct about the limitations of artificial intelligence, expressed strong reservations. In *Computer Power and Human Reason,* he did not content himself with critiquing AI techniques. He also delivered an impassioned rant against the moral basis of the entire discipline of artificial intelligence and the role of computers in society. Computers, Weizenbaum argued, encourage society to think it can solve problems that have been previously deemed unsolvable. Weizenbaum suggested that that was a misguided approach. Better to think of new ways to avoid having problems than to simply throw more processing power at them.

Weizenbaum had been worried ever since the creation of Eliza in 1965. Almost immediately after debuting the chatterbot, he had a falling-out with his collaborator, psychologist Kenneth Colby. On one level the dispute was over who should get academic credit for Eliza's creation, but on a deeper level the fight centered on the potential applications of Eliza-style technology. Colby, claimed Weizenbaum, believed that computer therapists could eventually play a role in the treatment of humans and would be invaluable by supplementing over-worked hospital staff.

That suggestion was inconceivable, Weizenbaum maintained. How could computers ever empathize with human suffering or truly form bonds with their patients? For Weizenbaum, admittedly not a psychologist, the act of therapy was one in which the human element was essential. Eliza could never understand the concept of love or hate.

The prospect of Eliza-like automatons entrusted with the mental health of a nation still appalls Weizenbaum today. The sacred doctor-patient relationship should not be given over to a mere machine. But how much responsibility should we cede to machines—or bots? Where does it stop? Who ultimately is responsible when they go awry?

Colby takes a much more instrumental approach to the psychoanalytic relationship. He's a firm believer that therapy can be reduced to a strictly defined algorithm. He dismisses Weizenbaum's fears as unrealistic and "moralistic." Although Weizenbaum never experimented further with natural language processing after the initial, dismaying success of Eliza, Colby has tinkered with Parry ever since. Today Colby is a professor of cognitive science and psychology at UCLA and, together with his son, is busy marketing Overcoming Depression, a program that uses conversational techniques to assist patients in dealing with their problems. Overcoming Depression is a direct descendant of Parry and Eliza. And although Colby does not outright recommend that his program be used for actual prescription of treatment, he still sees no problem with using programs like Overcoming Depression to supplement an overburdened health care system.

< >

Weizenbaum's fears find little counterpart in today's agent-crazed environment. One hears next to nothing from MIT's Nicholas Negroponte about the potential negative applications or implications of agents or bots. Computer scientists wave aside moral questions as if they were bugs in the system that are bound to be solved in the next upgrade. Frankenstein seems more special case than alarming specter.

Moral equivocation about artificial intelligence? It's a moot issue, declares MIT's Sherry Turkle. For years, Turkle has been specializing in the psychological aspects of human-computer interaction. In her most recent book, *Life on the Screen,* she recounts how every year she polled her students on the question of whether it would be wrong to replace humans with computers.

In 1990, for the first time, her students "saw nothing to debate." Pragmatism had replaced worrisome quibbles. If the programs had something of value to offer, that was all that mattered. There was no moral issue at hand.

Turkle is an unabashed postmodernist, a disciple of the school of theorists lumped together under the rubric "poststructuralist." She attributes this new pragmatism in part to a sea change in the overall culture, the transition from a modernist age in which people believed in ultimate values to a postmodern age in which everything is relative and there is no foundation. (This transition closely parallels changes in how the quest for artificial intelligence has been pursued—from knowledge representation to fuzzy logic.)

The postmodern era, claims Turkle, citing theorist Fredric Jameson, is characterized by "both a new 'depthlessness' and a decline in the felt authenticity of emotion"—a "waning of affect." That contemporary human beings are willing to trust their innermost thoughts to a computer program like Colby's Overcoming Depression, argues Turkle, is a sign of this depthlessness.

"In the emerging culture of simulation," writes Turkle, "they are happy, in other words, to take the program at interface value."

At first glance, Turkle's observation might seem to be one more part of the puzzle as to why bots are thriving on the Internet. People simply don't care. A culture raised on cartoons,

Arnold Schwarzenegger movies, and fast food simply can't bring itself to be critical. So what if bots aren't that smart? They're still really neat, and we like 'em.

But that's facile. People do care. Bot ethics, bot abuses, and bots gone bad are the Net's hot topics of the day. Mark V. Shaney and Serdar Argic pissed people off, not least because of the deception involved—they were programs passing themselves off as humans. When a mailbot malfunctions and starts spewing countless messages to a mailing list (or even worse, when an advertiser uses an automated script to send ads to mailing lists), people get angry. When a Web robot overloads a Web server and brings it crashing to a halt, people aren't assuaged by the fact that it has a name like "Wobot" or any other cute feature—they're furious at the waste of their time and money. When an IRC channel devoted to discussion of Bible issues is hit by a floodbot spouting sacrilege, the natives feel bothered and blasphemed against.

If anything, emotions are rising. Affect is not waning. It's waxing. And as bots get smarter, the fallout that their deeds generate will only intensify.

4

THE BOT WAY OF BEING

BEWARE THE WUMPUS

I am in a cave, connected to three other caves by short tunnels. I am armed only with a bow and a few arrows. Silence reigns, but danger awaits. Giant bats. Bottomless pits. And worse. Which way should I go? Cave 9? Or Cave 11? I take a deep breath, input the number "11," and press "Enter."
>You smell a wumpus<
The wumpus, I have been warned, is "kind of short, very heavy, has razor-sharp teeth, and legendary sucker feet . . . you wouldn't wanna meet him in a dark alley." I let fly with one of my arrows. I miss. Drat. I've got only one left, and I decide not to risk it. I choose another cave.
>Gulp—you've been eaten by a wumpus<
Time to start over.

Dating back to 1972, when it first appeared on the Dartmouth College Time-Sharing System, *Hunt the Wumpus* is one of the oldest of all interactive online computer games. The wumpus—that wily sharp-toothed beast—is a clear contender for the noble status of the world's first *gamebot*. It is also a tenacious inhabitant of the information wilderness. For years the wumpus has infiltrated corporate and academic computer networks, insinuating itself into the hearts and minds of the terminally logged in. Even today, although far outclassed by much gaudier action-adventure game challenges, the wumpus, like Eliza, survives on the World Wide Web. Old computer programs never die; cyberspace is their everlasting paradise.

Strictly speaking, the wumpus bears little resemblance to either a daemon or an Eliza-style bot. Indeed, the very idea that the wumpus might be a bot predecessor drove one computer programmer into a near fit of email apoplexy ("fer crissake, a daemon in *BASIC*? In *1972*?"). Only an ignorant, willfully reckless, hack journalist, he implied, would dare suggest such slander. The wumpus is not autonomous, has no intelligence to speak of, performs no useful service, and can hardly be said to have a single personality trait other than overweening voraciousness. The wumpus isn't even a glorified subroutine. After each decision a player makes in *Hunt the Wumpus,* a random number generator determines whether the wumpus eats, steps on, or bumps the player; whether it dodges the arrow or is killed by it.

Humble though it may be, the cave-dwelling wumpus owns a spot in bot history, if only in tribute to its inspirational example. Wumpus offspring have spread to the far corners of the Net.

In 1988, shortly after Finnish computer researcher Jarkko Oikarinen programmed the chat application that made the Internet Relay Chat network possible, he decided to make the source code to the software available to all. He hoped that interested parties would seize the opportunity to set up their own IRC *servers*—computers capable of accepting IRC connections from other computers—and thus spread the IRC network around the globe. He envisioned a world of gossip mavens linked together in thousands of online chat rooms (which he dubbed channels), freely discussing whatever topic they cared to blather about.

Along with the source code, Oikarinen included a template for creating what he called an automaton—an independent piece of code that could be summoned by IRC users to any IRC channel and adapted to provide various services. He described it as a "primitive artificial intelligence" that could recognize words in an ongoing IRC conversation and participate in the discussion with a set of predefined answers. Early IRC adopters grabbed the automaton code and immediately warped it for their own purposes. One Finnish colleague of Oikarinen's created a chatterbot named Puppe, "a simulation of an early IRC nerd" who spoke Finnish and is reported to have had several moods, depending on his level of virtual intoxication. One American adapted the code into a bartender bot that mixed virtual drinks for channel lushes. And another American, Greg Lindahl, programmed the gm bot; this gamemaster bot could conduct a game of *Hunt the Wumpus* in any given IRC channel.

That gm bot inspired Jim Aspnes at Carnegie Mellon University to write TinyMUD, the spawning ground of Mauldin's Maas-Neotek chatterbots. The wumpus is the direct ancestor of Julia. In fact, the wumpus, one could argue, helped shape the evolution of three of the greatest bot lineages—gamebots, MUD-based chatterbots, and IRCbots.

What gave the wumpus so powerful an impact? How could a few lines of Basic make such a lasting impression on so many successive generations of the computer literate?

The wumpus adds a major ingredient to the bot mix: environment. The sucker-footed beast is one of the first *situated* interactive characters in a computer-created world. It has a habitat—a dodecahedral lacework of dark, dank caves. It navigates—or, to be precise, gives the appearance of navigating—amid gaping chasms and bat colonies.

The player enters into that environment and battles with the wumpus monster on its own turf. That turf, of course, is imaginary—there is no physical reality to it, nor even a graphic depiction of it—but even imaginary turf is better than no turf at all. Given just the sketchiest of hints from a bare-bones text interface, the human mind is capable of dreaming up a landscape as vivid and rich in detail as that produced by the most talented Hollywood director. All it needs is a starting point, and that's what the *Hunt the Wumpus* game provided. That's what the *Hunt the Wumpus* game *had* to provide, for there is no fearful wumpus without the dark mystery of the caves, just as the caves are meaningless without the wumpus to make them dangerous.

By contrast, in the case of Eliza, there is no there, there. No office with diplomas adorning the walls. No chair, no couch, not even a pencil and pad with which the therapist can pretend to scribble obscure notes or doodle the hours away. Eliza exists as a free-floating wraith-doctor, substantiated only as electronic pulses coursing through computer hardware.

Situating a character in a world defined by a few adjectives and nouns is an old human trick. Characters drive all drama, whether in film or fiction, chants or comic books. And no character exists in a vacuum. Characters appear in scenes. They require a fully realized world to give context to their words and deeds.

The same goes for bots, with a bonus. A well-executed, well-thought-out bot environment makes a bot all the more believable. But most important, it allows meaningful interactivity to flourish. That lure of interactivity makes up for all manner of bot inadequacies. Interactivity gives us—humans, users, players—a role. A person entering into a bot world is no mere spectator confined to a seat at a movie theater, no passive reader limited to turning pages in a book. That person has leapt onto the screen, into the narrative, is confronting the characters directly, seeing them for himself or herself. We don't just read about someone hunting a wumpus. We hunt.

Not all the bots running free on the Net are situated in environments or, for that matter, are particularly interactive. Cancelbots, IRC warbots, and Web robots are just a few examples of programs that are loosely labeled "bots" but that still go about their business without personality and without "place." But even if the bot has no inherent personality, it may well have a name: Dave the Resurrector, Lazarus, Scooter, Gargoyle. Personification comes naturally—we are unable to content ourselves with the raw data of the human-computer interface. We want more, require more.

In games, chat rooms, MUDs, and other computer environments, hackers and hobbyists have been fascinated by the possibilities of interactive characters wherever technology has made them possible. But it isn't only a desire to tell a story that pushes bot character development. The wumpus and its kin are ample testimony to the longing of humanity to seed whatever environment it encounters—silicon, biological, physical, spiritual—with creatures of its own fancy.

The anthropomorphic urge—our human propensity to endow everything in our world, animate or inanimate, with personality—is impossible to suppress. Bots are as much a product of that urge as they are entities of their own accord, as much creations of our minds as they are clever conglomerations of ones and zeros. True botness cannot be reduced to drab sequences of Lisp or lists of functions. Bots are products of the human imagination, the result of our relentless need for companionship, our fear of being alone in a great big universe. Bots exist for the same reasons that the ancient Greeks warped constellations of stars into the shapes of great bears or scorpions. They make our lives richer, more interesting, and more entertaining.

THAT BASTARD-BOT

> *I am at one end of a vast hall stretching forward out of sight to the west. To my right, a wide stone staircase leads downward. Wisps of white mist sway to and fro, almost as if alive. A cold wind blows. Somewhere around here, I know, is a heavy gold nugget. I want it. But wait! What's that sound?*
> >A little dwarf just walked around the corner, saw you, threw a little axe at you which missed, cursed, and ran away.<

Foulmouthed dwarfs, giant snakes, cheering bands of friendly elves—chances are, if you were into computers in the 1970s, at one time or another you became addicted to the game *Adventure*, the first so-called interactive fiction computer game.

As Dartmouth students whiled away the hours calculating wumpus extinction probabilities, a hundred or so miles south, in Cambridge, Massachusetts, at a small scientific company named Bolt, Beranek, and Newman (BBN), professional engineers discovered a far more engrossing way to ruin their productivity: *Dungeons & Dragons*. A role-playing fantasy-adventure game in which players toss dice to decide their likelihood of surviving armed battle with goblins, trolls, and a host of other magical foes, *Dungeons & Dragons* caught the imagination of a generation that had just read J. R. R. Tolkien's *Lord of the Rings* trilogy and hungered to take a broadsword into battle and hack away.

BBN is notable in its own right as the enterprise that received the first Defense Department grant to construct the Arpanet, the precursor to today's Internet. For years, BBN had been cherry-picking the best and brightest graduates from MIT. Will Crowther was one such MIT alumnus, a former physicist who had recently gravitated to computer science. A superb coder, skilled rock climber, and dedicated cave explorer, Crowther combined two of his passions in 1976 and wrote *Adventure,* a simplified computer version of *Dungeons & Dragons* set in an underground maze of caves.

After Don Wood, a graduate student at Stanford, expanded the scope of the game and embellished it with Middle Earth imagery, *Adventure* became an icon of the computer generation. It forever changed the way people thought about computers, foreshadowing the doom of static paper-and-dice games like *Dungeons & Dragons.* Why roll dice when you can have a random number generator do it for you, behind the scenes, and thus allow you to avoid facing up to the dire reality that all of fate is chance?

Adventure poses a sequence of puzzles. How do you get past that giant snake? When are you supposed to chant the magic word "xyzzy"? Where do you go from the Hall of the Mountain King? How do you avoid getting lost in the forest? And what do you have to do to keep from getting killed by those damn dwarfs?

Seven dwarfs in all, armed with daggers and axes, move from cave to cave in *Adventure,* progressing according to the topology of the labyrinth. Technically, the dwarfs, like the wumpus, do not satisfy the requirements for being considered autonomous processes. The whole game of *Adventure* is a single process, or program in action, and nothing happens until you, the player, make a move. At that point, the game updates its internal data structures, and the various dwarfs advance one step forward.

True autonomy is rare in cyberspace, and for most practical purposes it is irrelevant, a quibble. Technicalities mean next to nothing when the human imagination is involved. Believability comes cheap. Two or three lines of text made *Adventure's* dwarfs as real as was necessary. In early 1977, *Adventure* swept the Arpanet. Today you can hardly turn a corner in cyberspace without having an ax thrown at your head. *Adventure*—and its worthy successor, *Zork*—paved the way for a vast world of fantasy-adventure MUDs and the even more wide-ranging universe of gamebots.

Zork's creators, a quartet of computer science graduate students at MIT who quickly became bored with *Adventure* and decided they could do better, willingly submerged themselves in the muddy waters where technical reality and make-believe collide. *Zork*, with its menagerie of trolls, wizards, monsters, gnomes, princesses, and thieves, upped the ante on *Adventure*. The thief, in particular, made *Adventure's* dwarfs, their sharp little knives notwithstanding, positively innocuous.

"The basic idea was that we had seen the really cheesy demon in *Adventure*," remembers David Lebling, one of *Zork's* original authors, "the nasty little dwarf that appeared and threw daggers at you."

"We wanted something that acted intelligent," says Lebling, who now works as a programmer for Avid Systems, a company specializing in high-end digital editing machines. "The thing that got the most work in that regard was the thief. He was really pretty sadistic—he would pick your pocket, make cutting remarks about you. A lot of loving care went into making him as sadistic as possible."

Before working on *Zork*, Lebling had cowritten a computer game called *Maze*. In *Maze*, multiple players wandered about, attempting to find the exit and shooting at each other along the way. If no other humans were available to play the game, a

human player could activate several robot players that specialized in hiding in corners, waiting for a good shot, and then
running away. *Maze* was written in 1973, so Lebling has to
rank as one of the primordial gamebot developers. Yet here he
is now, loosely flinging around definitional terms that do not
jibe with bot reality.

A cheesy "demon"? Lebling and his fellow *Zork* authors
had offices in the same building as MIT's Artificial Intelligence
Laboratory, where "demons"—in the sense of both Corbato's
daemon and the even earlier Pandemonium AI demons—were
a hot, ongoing topic of research. Lebling knows quite well
what the word *demon* means in a computing sense, and when
pressed, he admits that *Zork*'s characters did not fit the description. But he uses the word anyway, in published articles
and interviews.

Lebling has fallen victim to the Eliza effect. He is thinking
about his code creations as actual characters. He is a believer.
As a lord of *Zork*'s creation, he knows better and will confess
the truth when directly pressed. But programmers, perhaps
even more than gameplayers, yearn to imagine that their code
adds up to something more than just another number. They
want to believe. And they know, or at least the smart and honest ones of them will admit, that making a bot or character act
intelligent is more than half the battle. Alan Turing laid down
the party line. A bot doesn't need to *be* intelligent. It just needs
to be able to fake it.

Who cares, anyway, if the bottom line on a wumpus or a
dwarf or a troll is that it is far closer to a digital cardboard stick
figure than to an autonomous intelligent program? Just ask a
Hunt the Wumpus player who has been "gulped" if the wumpus
really exists. Or better yet, listen to some of the comments
posted to a Usenet newsgroup by players of the popular shoot-
'em-up game *Descent*—twenty-five years after the birth of the

wumpus—as they ponder how to deal with a great-great-great-great-grandson of *Zork's* thief, the dreaded *Descent* thiefbot.

> I have *chased* that little $##@% all over the mine (blasting away feverishly—with what I had left—in a fit of frustration) and not managed to kill his spiny green hide. . . .

> He's tough . . . the little turd. . . .

> I really hate that bastard-bot. . . .

> When the thieving bastard tries to sneak up on you turn around and blast him as he runs away down the hall!

Descent's basic plot conceit is that rebellious work robots have taken over a string of mining colonies and space stations. What's worse, they've gained intelligence to the point of being able to modify their pulse rock cutters and argon welding lasers into dangerous weapons. The thiefbot is a vile robotically self-induced mutation—an annoying little speedster who sneaks up behind your virtual back, steals all your weapons and ammunition, and then peels away, usually with you in hot and fruitless pursuit.

Humans against bots! *Descent* is rife with subtexts providing ample fodder for the discussion of gamebot psychology. The rebellious robots are Frankenstein monsters, truculent reproaches to human mastery. Their supposed ability to mutate into new forms suggests malignant viruses at loose in the dark corners of cyberspace. But beyond the idiosyncrasies of the game itself, the online agonizing of *Descent* players admirably reflects all gamebot-human relations. A similar breathless personification takes place in any popular computer game—especially an action-adventure or fantasy role-playing one.

Only recently have modern computer game developers begun to recognize the built-in attraction that a well-developed game character exerts on players. During the eighties, text-based games like *Adventure* and *Zork*—which emphasized story lines and characters—were gradually eclipsed as personal computers increased in power. Game developers devoted the bulk of their resources to taking advantage of processing power, memory, and graphics capabilities for ever more cutting-edge sound, video, and action.

But today the best developers race to integrate more lifelike, or at least more compelling, characters into their story lines. A good gamebot is a real selling point. So now the marketing campaigns for the newest games emphasize the incorporation of the latest advances in artificial intelligence. Genetic algorithm breeding is supposed to permit gamebots to learn human player strategy and evolve into more fearsome competitors. Each tiny advance in natural language processing is speedily adopted by developers hungry to improve their characters' speech capabilities. The gaming industry is a multi-billion-dollar-a-year behemoth that may ultimately be far more influential in pushing the development of functional AI strategies than the combined might of academic research. Interestingly, particularly considering how problematic the achievement of true AI really is, these developers don't need state-of-the-art AI techniques to create winning characters. They may be wasting their time. The general public is ready to believe in just about anything, as long as it is cleverly constructed. From *Hunt the Wumpus* to *Descent,* and in a hundred gameworlds in between, bots benefit mightily from the innate human tendency to personify. A good game designer or bot designer should care less about "intelligence" and "reality" than about taking advantage of our simple willingness to believe.

That simple willingness is often and paradoxically expressed as the ability to suspend disbelief. Encouraging the suspension of disbelief is one of the oldest tricks in the book for storytellers and for that modern incarnation of the storyteller, the graphic designer. Graphic designers—and by extension, those who would design agents or bots in the form of believable characters—would do well to study the example set by the first interactive computer fiction characters. Too much reality can spoil a good thing, especially if done badly. Just a little touch of character, sketched on the canvas with a few strokes of the brush, and just a few sentences of text can achieve astonishing results. And even though designer emphasis on interactive game characters sank into lamentable obscurity in the eighties, overshadowed by ever more spectacular computer pyrotechnics, that lesson has been heard loud and clear on the Net.

THE DWARFS OF CYBERSPACE

I am in a place called the Patpong Go-Go Bar (also known as room #10522 in LambdaMOO). Stevie Wonder's "Part-Time Lover" blares from speakers on either side of the bar. On one side of the room is a stage with six dancing go-go girls; on the other side, a stage with six dancing go-go boys. I see a lineup of prostitutes, and Newt Gingrich. I type the command "pay Newt."

>Newt Gingrich says, "Thanks. Now I will take you into a corner and give you your orgasm privately."<

>Newt Gingrich opens your shirt and runs a hand down your chest.<

>Newt Gingrich lowers his lips to your nipple and sucks on it.<

>Newt Gingrich moves his other hand up the inside of your leg and rubs your crotch.<

I run away, screaming in horror. Some virtual realities are better left unimagined. I'm a big bot fan, but Newt Gingrich the prostibot, a

product of the warped mind of author and MOO experimenter
Cleo Odzer, is one bot I don't ever care to meet again.

MUDs picked up where interactive games like *Adventure* and *Zork* left off. In fact, the very first MUD, created by University of Essex at London graduate students Roy Trubshaw and Richard Bartle in 1979, was directly inspired by the two trailblazing fantasy games. The original meaning of the acronym "MUD," now generally used as an abbreviation for "multiuser domain," was actually "multiuser dungeon." Bartle and Trubshaw had been avid players of the games, but they saw room for improvement. MUDs offered the prospect of multiplayer participation. What's more, Bartle felt limited by the one-track mind of games like *Adventure* and *Zork*. He wanted a more open-ended environment.

Bots appeared on EssexMUD almost immediately, at first in primarily utilitarian roles, but later as more ambitious figures in a wild tapestry of heroes and fools.

EssexMUD predated the widespread Internet connectivity we enjoy today. In 1979 most EssexMUDders could access the MUD only through a direct dialup phone connection. And only a limited number of people could be connected to the MUD at once. To ensure room for new MUD visitors, Bartle wrote a program that disconnected players whose accounts had been inactive for a given period of time. Get up, go to the bathroom, and you might lose your place on the MUD.

But clever users immediately found a way around this facile attempt to get rid of them. They wrote simple placekeeping programs that could periodically take some form of action and thus keep the account alive. Typically, the program could be as mindless as one that would utter "heh" every five minutes or so. In the forgiving text environment of a MUD, the word *heh* is almost always contextually accurate, so no one would be wise to these primitive chatterbot games.

"They weren't all that sophisticated," remembered Bartle, who eventually gained a PhD in artificial intelligence but now makes his living as a full-time MUD programmer. "Initially these programs just did things like output 'hmm,' 'really,' or 'hehehe' periodically. Amazingly, people would still carry out conversations with them. They did get a bit more sophisticated later on, but since they were written by hackers rather than AI students, they never really got up to even Eliza level. . . . If they could trick someone into talking to them for a minute or so, that was good enough for their authors."

Place-keeping bots, like humans and like the Julias to come, logged in from "outside" the MUD. Their actual code would reside where the human running the bot had an account, or access to a networked computer. Such bots are referred to as external bots, and some purists consider such externality to be a required characteristic for true bots. But external bots are often grouped together with so-called internal bots—a category that includes the classic gamebot, or computer-generated character. An internal bot's code is part of the MUD or game itself. "Puppets," "mobiles," and "nonplayer characters" are all terms used to describe various types of internal MUDbots.

Bartle's fertile imagination filled his MUDs with a host of internal MUDbots, some of which were classic examples of personification, like the normally placid tree nymph who could become a fierce fighter when provoked (and, according to Bartle, could be easily vanquished by being set afire). But Bartle draws a line between his internal creatures, which he calls mobiles, and bots. Bartle considers a wisp of breeze floating through a MUDroom to be a mobile. As he observes, to include gusts of air is to seriously stretch the definition of bots.

The MUDs that Bartle is writing these days achieve astonishing levels of complexity with his elaborate uses of internal MUDbots. Listen to him describe via email the dwarf colony in a MUD he created for the commercial multiuser online gaming system Kesmai Aries, in 1996:

> I have a colony of about 50 dwarfs, who live in an underground realm. When a player opens the gateway to the realm, there's a crack of thunder. This tips off the dwarfs that there's someone out there, and one of them is charged with the job of closing the gateway. To do this, she has to figure out a route from her current location to the gateway, opening any doors which are in the way (she has a key, but if someone has removed it she'll try to find another), until she gets to the gateway, which she then locks. Meanwhile, the other dwarfs (those which aren't on guard duty, asleep, or which are children) will make their way generally towards the dwarfen hall which lies at the bottom of the entrance. If a player enters, the dwarfs will not attack, being curious yet cautious about them. However, if a player picks up anything belonging to the dwarfs, they WILL attack. What's more they'll do so in a coordinated fashion: if a dwarf sees another dwarf under attack, then they'll join in on the dwarf's side. Eventually, players will realise this, and leave all dwarfs alone except those dwarfs which guard the route to the treasure chamber. One of the guard dwarfs carries a shortsword which is particularly effective against dwarfs, and gives its wielder a fairly decent chance of killing the dwarf king (who is able to cast spells, and does so reasonably intelligently depending on his current situation). If the players are intent on butchering the rest of the dwarfs, they will probably be able to manage it using the shortsword and knocking back wafers (restorative of stamina). However, the smartest players have waited

for the dwarf whose job it is to close the gateway, killed her when she arrived, and left the gateway open. The dwarfs (particularly the younger ones) who have been living in the citadel all their lives are keen to explore, and will venture out into the rest of the land. Since they need to negotiate the tunnels of a tin mine to get out, and there's a great cistern full of water which can be released by turning a valve, it's quite possible for players to wait until a good many of the dwarfs have left the citadel, then turn the valve and drown them. This doesn't score so many points, but it's a lot less risky than hacking each one individually!

Clearly, the dwarfs of cyberspace have come a long way since their days throwing axes in *Adventure*. And so have MUDs. EssexMUD begat AberMUD. AberMUD begat DikuMUD and LPMUD. Then came UberMUD and UnterMUD and Tiny-MUD. Not to mention MUCKs, MUSHes, MUSEs, and MOOs. TinyMUCKs, TinyMUSHes, TinyMUSEs, and TinyFUGUEs. And countless other variations on a multiuser theme.

To differing extents, bots, both external and internal, have flourished throughout this expanding MUDverse. But there are some limitations. Not all MUD programming languages are as friendly to the creation of bots as TinyMUD and MOO are. Some MUDs even have rules against allowing any outside bots, other than wizard-approved internal characters. And in general, each MUD variation is a separate ecology with its own physical laws, which militates against the widespread dispersal of successful bots. A TinyMUD bot cannot exist in a MOO without substantial reprogramming. In evolutionary terms, the MUD universe is more like a grouping of isolated South Pacific atolls than a unified continental landmass.

MUD evolution itself is threatening the bot ecosystem. Paradoxically, even as state-of-the-art computer games are in some ways becoming more MUD-like, with an increased emphasis on story line and character development and on multiplayer capabilities, the cutting edge of MUD development is moving along the same trajectory that games did in the 1980s. First, the World Wide Web added graphic and audio capabilities to the Internet's text-only user interface. Now, more ambitious technological advances propose to transform the Web from a two-dimensional chain of linked documents into an interactive three-dimensional world full of sound and visual fury. These virtual worlds, or Web-based MOOs (WOOs), require powerful computers for satisfactory results, but they are the focus of steadily increasing user and investor interest.

That may not necessarily be good news for bots. When you strip away the protective cover of text and enter a fully 3-D realm of real-time interaction, it becomes more difficult for bots to garner that necessary suspension of disbelief. It is far harder to model live-action physical movement and integrate real-time voice recognition and speech generation into a believable character than it is to hack up a compelling text-based chatterbot. Avatars—computer-generated online graphic images that a human can choose to depict himself or herself (pharaonic wigs, baboon faces, abstract shapes) in an online environment—are clumsy and awkward. In these new animated 3-D worlds, bots can't get away with as much as they used to.

Characters like the wumpus, the dwarfs and thieves of *Zork* and *Adventure,* and the myriads of bots wandering everywhere through MUDs aren't planned, designed by committee, or aimed at a particular marketplace. They are examples of the grassroots energy, the bottom-up experimentation, that propels the bot movement forward. But in this new, ever more technically complicated world, can such energy persist? Can the hobbyists who have pushed so much of bot development up to this point take bots to the next level?

Or is bot evolution poised for a new stage altogether? Will commercial exploitation become the new engine of bot development? Game developers aren't the only ones interested in coming up with compelling, seemingly intelligent computer-generated characters. As interactive computer culture has spread beyond the insular worlds of geeks and academia and into the mainstream, the arbiters of commercial culture have become steadily more interested in the possibilities of botlike characters. The potential benefits of their use—as interactive librarians, animated cartoons, movie spin-offs—bear the scent of a gold miners' rush, sweaty, exuberant, and based more on greed than on prospector surveys. Bots have big-time product potential, if someone can figure out how to make them work.

THINGS THAT THINK FOR PEOPLE WHO DON'T

A little dog is wagging its tail and telling me to "click on the door and sign in." His name is Rover Retriever ("but you can just call me Rover"). Most of the time, Rover communicates via cartoon-style speech balloons, but he audibly whines when I don't follow his instructions correctly, and he pants while taking care of business.

I'm sick of dogs showing me around cyberspace. Lately the Net has become like a leashfree park on Saturday, nothing but hounds—WebDoggie, Fido, Rover—running around, sniffing, and barking. I decide to change my personal guide and go with Orby, "the greatest globe you'll ever meet." Orby is "worldly, lovable, carefree, and whirling with energy." His favorite food is star fruit, and his birthplace is the Milky Way. He's a blue and green globe with stick figure arms and legs and an aquiline nose. Click on him and he does a little pirouette.

I play around for a few minutes, and then I delete every trace of Rover, Orby, and all their silly friends from my hard drive. I do not need pirouetting anthropomorphized globes in my life. There is

a limit to how personified I want my computer interface to be, and
these creatures, the Friends of Bob, have crossed over the line.

Given the great fanfare, the promise of a hundred-million-
dollar marketing campaign, and the high-profile leadership
of Melinda French Gates, the wife of Microsoft chief executive
officer Bill Gates, one would have imagined there to be no way
that Microsoft Bob could go wrong. Promoted as the future of
computing at its January 1995 debut, trumpeted as a revolution in computer user-friendliness, Bob was to be a shining
example of a new paradigm in human-computer interaction—the so-called social interface.

Bob consists of a collection of bare-bones programs, including a no-nonsense word processor, a calendar, a check-
book, an address book, and an email program. Aimed at the
first-time computer user, Bob is designed to make interacting
with these various programs as easy and unthreatening as possible.

The Friends of Bob are the focal points of interaction.
(There is no actual "Bob" in Microsoft Bob, a fact that may
have encouraged the computer press to let their sarcasm me-
ters go off the scale.) The Friends of Bob include fourteen "personal guides," each of which has a different personality. In
addition to the aforementioned Orby and Rover, there are
Baudelaire the moody gargoyle, Scuzz the sewer rat, and an
oddball assortment of other goofy creatures. The guides express their personalities through pithy one-liners and brief animated movement—when surprised, Rover may utter a phrase
like "Land Sakes!" and waggle his ears.

The guides offer a series of choices for navigational tools,
but Bob includes no manual. It is the personal guides or nothing (although for those who absolutely can't abide neurotic

"gryphons" or importuning animals, one guide takes the form of a no-nonsense wall-mounted speaker).

Bob is based on several years of research by two Stanford professors who consulted for Microsoft on the design of the program. The professors, Byron Reeves and Clifford Nass, cochair the Social Responses to Computer Technology PhD program at Stanford. For half a decade, the program has been conducting psychological tests to determine how people respond to computers.

According to Reeves and Nass, the results of the tests showed that people treated the computer as a "social actor." In other words, if praised by the computer, a person tends to evaluate the computer more highly. If criticized by the computer, the person is less favorable. Reeves and Nass found that test subjects responded positively to computers that were polite or prone to flattery or whose "style" matched the subjects' own. Reeves and Nass concluded that human-computer interaction would proceed best if conducted along lines that recognized the social actor status of the computer. Designers should think in terms of making the interaction as pleasant as possible or as geared to the person's personality as possible. Bob is the first major commercial test case of their theories.

"This is a revolutionary change in the way people use computers," said Nass, as quoted in numerous press releases at the time of Bob's debut. Bob, said Nass, represented the "third wave of computer interfaces." First came the text-based interface. Then the graphical interface. And now, finally, the social interface.

The Friends of Bob can be considered extremely simplistic intelligent agents or primitive bots. Either way, they mark one of the first big budget attempts to move the world of personified computer programs out of the game/MUD ghetto and into the operating-system marketplace. But if Bob was a test case for the potential of intelligent agents married to a personified interface, or for the prospects of a day when there will be a bot on every desktop, then Bob flunked.

Failure really is too mild a word to describe the public's response to Bob. Bob bombed. Savaged by the critics and rejected by consumers, Bob was pilloried as condescending, patronizing, and an insult to the intelligence of a two-year-old. Bob became exhibit A in the case against the anthropomorphized interface, delicious grist for the mill of those specialists in human-computer interaction who believe that the personification of tools is a terrible idea, that bots should remain confined to their gameboxes and isolated from any important work role.

Perhaps the most articulate of those critics is Ben Shneiderman, a computer scientist at the University of Maryland at Baltimore. Shneiderman, a man of razor-sharp intellect who lays out his arguments with the panache and precision of a classically trained debater, takes pleasure in ripping apart the rhetoric of agent technology. His central point is that an anthropomorphic interface is an inefficient interface, more of an obstacle to work than an aide, the relic of millennia of animist superstition. Ben is anti-bot.

"The designers are infatuated with the anthropomorphic scenario, as they have been for thousands of years," says Shneiderman. "For those who built stone idols or voodoo dolls or the golem or Frankenstein, it's long been a dream. . . . But no mature technology resembles human form. Automobiles don't run with legs, and planes don't flap their wings like birds. As an early inspiration the anthropomorphic scenario is a good idea, but if you stick to it, you're going to miss the grand discoveries of effective technology."

Shneiderman blames the overweening ambition of the artificial intelligence community for holding back the real work of human-computer interaction. When I offer him the standard answer of the computer industry to criticism, the defense that Bob is "just a bad implementation" and that advances in processing power, AI, voice recognition, and other related technologies will no doubt solve current problems, he snorts.

"They've been telling me that for thirty years," he says.

Shneiderman concedes, grudgingly, that natural language processing programmers and personified interfaces have been successful in game settings or as context-sensitive help menus. But what really gets his goat is the emphasis on autonomy, the "indirect management" that is supposed to be the biggest selling point of agent and bot servants. Indirect management, argues Shneiderman, is a recipe for disaster, a guarantee of inefficiency.

"You need to know that if you issue a command, you are going to get an exact and repeatable operation," says Shneiderman. As an example, he cites the task of moving computer files from one directory to another. With the help of a mouse and a drag-and-drop interface, the job can be accomplished cleanly and efficiently with the flick of a wrist—click, drag, release. But imagine if instead of dealing with a mouse cursor, you were speaking in a natural language to an intelligent agent. You couldn't just say, "Move these files over there." You'd have to say, "Move the files in directory C:\winword\bots\ai to the A drive." And as soon as you start speaking in tongues, the potential for confusion becomes dramatic. You don't want fuzzy logic when you're working. You want the right result, in the least amount of time. Shneiderman firmly believes that users will reject indirect management. They'd rather have control. Menus. Windows. Direct impact.

"I've got to go with the successes," says Shneiderman. "Human visual capability is enormous, and the computer is an ideal medium for visual display. The data rates you can get by pointing, clicking, and dragging are a hundred to a thousand times faster than natural language typing or voice."

Furthermore, argues Shneiderman, thinking of an agent or the human-computer interface as a person or even as an animal is a plunge downward into murky waters of accountability and responsibility. If one allows oneself to anthropomorphize a dumb tool, then one may begin to think of that tool as being responsible for its actions. So what happens if a bot runs amok? Do you blame the bot? or yourself? or the original programmer? How much easier is it to distance yourself from personal accountability if you can blame "someone else"?

Agents, concludes Shneiderman, are crutches that don't work, mere invitations to mediocrity. They are "things that think for people who don't."

MORE CATS AND DOGS

I move my mouse cursor across the screen. Jester pokes her head out of a wicker basket. I double-click the basket, then swerve the cursor up across the monitor to the right. The fuzzy little kitten immediately chases it across the screen. I halt the cursor. Jester sniffs at it suspiciously, circles around it, swats it with her paw. I move sharply to the left. She does a back flip and follows. I place the cursor on the kitten's back and start rubbing. After a few seconds Jester arches her back and begins to purr.

A bottle of milk, a ball of yarn, a hunk of cheese, and a mousehole are stashed on a bookshelf to the left of the monitor screen. I click on the bottle and drag it over to Jester, who rolls on her back, grabs it with all four paws, and starts sucking. I know that if I fail to feed the kitten regularly, she will become progressively more listless and will eventually refuse to play with me.

I click on the upper right-hand corner of the window to close the program. A new window pops up: "Adopt Now? or Abandon Us?"

Being a hardened interactive experimenter and having trained myself to withstand the seductive tricks of interactive interfaces, I have no problem abandoning Jester the kitten to an unknown fate. But I do feel a little twinge. The kitten is believable, a superb example of how to model cat behavior in a software simulation. The essence of kittitude has been captured.

Catz is a screensaver game. Jester and her siblings aren't pretending to be interface agents—their purpose is merely to entertain, and they do their job well. They are certainly believable enough that at PF.Magic, the small San Francisco–based software company that developed *Catz* and its companion screensaver game *Dogz,* there has been some discussion as to what should happen to cats whose owners refused to feed them. Allowing cats to starve to death, decided PF.Magic, was unlikely to be a sound marketing strategy. Thus the listlessness.

Catz offers a taste of a workable social interface. And soon to follow it into the marketplace came another: one of the first commercial intelligent agent offerings, AgentWare's AutoNomy.

The AutoNomy agent is a tool for searching the World Wide Web for documents relevant to particular search areas. The underlying technology is similar to that of the Web robots that build up the indexes for the Web's big search engine services. But those Web robots work behind the scenes, out of sight of the millions of Web surfers who flock to the big search engine sites every day. AutoNomy is your own personal Web robot, living on your computer, waiting to do your specific bidding.

The AutoNomy interface is a dog. Actually, it's several cartoon dogs, generic mutts with floppy brown ears. On the opening AutoNomy screen, three of them sit patiently, eyes and heads swiveling to follow any mouse cursor arrow movement across the screen. Every so often one howls in boredom. The dogs, which have no names of their own to start with, need to be trained before they can be used. Training occurs by inputting a few natural language sentences into a form.

I name one dog BotHunter and input a few lines from a story I have written about bots. The dogbot then starts sending out HTML *calls* to retrieve pages that it thinks are relevant. Using an impressively effective system of text analysis, BotHunter quickly finds some good hits—Web pages that I agree are relevant to my search query (though nothing that I haven't already found the old-fashioned way, by trudging to search engine and Internet directory sites myself). As BotHunter searches, small animations run on the screen informing me as to his current status.

BotHunter can display eight states of behavior. My favorite is the one where he sits up on his hindquarters, holding out his paws, "begging to be let in." This signals that he is issuing a request to access a page that for some reason (such as bandwidth congestion or server problems) isn't being acknowledged. Other winners include BotHunter with his snout to the ground, digging furiously—"on the scent of something." A third animation has the dog with a bone in his mouth—"he brought back something relevant."

My BotHunter has little of the appeal of Jester the kitten. He isn't very cuddly or compellingly doglike, and he will not grow older or starve. He does work pretty well, though, better than I had expected, given my experience with other off-the-shelf search robots. And it isn't at all difficult to see how BotHunter and Jester could be integrated. Someday very soon, I am sure, I'll be able to click on some frolicking kitten right off my screensaver and watch her instantaneously kick aside her ball of yarn, morph into Bagheera the Web-Hunting Black Panther, and ask for instructions.

BotHunter and Jester demonstrate that modeling animal behavior is a great deal easier than modeling humans. Dogs are a popular icon for bots and agents precisely because we don't expect them to be as intelligent as humans. A wagging tail here, an arched back there—we're sold.

But even more significantly, BotHunter and Jester are counterexamples to the dismal failure of Microsoft Bob. People will respond positively to anthropomorphized interfaces, if done correctly. There is room for cross-fertilization between the grassroots energy of wild bots on the Net and the ceaselessly-searching-for-yet-another-consumer-sensation push of the marketplace.

And indeed, in the spring of 1997, the release of Microsoft's Office 97 suite of software applications demonstrated a great leap forward from the days of Bob. Office 97 comes equipped with cartoon "Office Assistants" (dubbed "toon-zombies" by some less-than-impressed critics) that offer an interactive help interface. The assistants are the clear sons and daughters of Bob, but they are less obtrusive and display some fairly impressive natural language processing abilities. And, most important, they can be turned off.

Microsoft is apparently learning that the critical area of interface design is not dependent on the brute force of processing power or better AI tools. Believability makes the bot. As do sensitivity to and respect for the needs of real live humans. On every front, Microsoft Bob failed.

"Bob sucked," says Abbe Don, one of the leading lights of modern human-computer interface design. "It was a childlike interface for adults. Bob was incredibly unsophisticated. That doesn't mean anthropomorphism should be thrown out the door. It means bad anthropomorphism should be thrown out the door."

The San Francisco–based Don, half artist and half programmer, has an impressive résumé of digital projects. One of her specialties is a subfield of design that deals with what she likes to call "characters in the interface." In the late eighties, she collaborated on a project at Apple to design personified computer guides that led people through a multimedia project involving a detailed historical database—complete with pictures and audio. She has also done interface projects for the Baby Bell telephone company, Ameritech, and spent several years working on a project to demonstrate the potential of a multimedia programming language jointly developed by Apple and IBM (and later scrapped, before ever reaching fruition). In a young field, she's one of the most respected practitioners.

Don believes that serious interface designers have resisted attempting to create fully anthropomorphic agents because they are afraid that modeling a "complex humanlike personality" is an impossible task. But that kind of modeling is done all the time in the art and entertainment world, she observes.

"There are two schools of thought on anthropomorphism," says Don. "One is what I would consider quite traditional, and it grew out of Joseph Weizenbaum and his experiments with Eliza. It said, Do not anthropomorphize the interface. It's misleading. . . . [Some people] have gone so far as to say it's unethical. And then there are those of us who say, Now, wait a minute. We have this whole field of the humanities and theater and film and comic books and all kinds of other representational arts where we represent humanlike traits through characters, and the audience is quite sophisticated about understanding that they are only representations, or metaphors. Most people do not mistake them for real people."

In fact, argues Don, the real mistake is to try to make characters look like real people. Abstract representations are more attractive. "The less realistic the character," says Don, "the more people project onto it." Just as text-based bots are easier to believe in than 3-D avatars, so are abstract representations of characters easier to relate to.

Abstract representations have resulted in some of the most successful characters in history, says Don, citing Mickey Mouse, a cartoon figure comprising little more than a circle and two ears. Abstract representations respect a person's intelligence, giving the imagination a chance to run free. Abstract representations are more believable, precisely because they do not attempt to model reality perfectly.

That quality of compelling belief, argues Don, is what a successful interface requires. Recall Socrates' daemon—it had no physical representation at all but still served admirably as intermediary, however one imagined it. To borrow a phrase coined by Pattie Maes, the director of the Autonomous Agents Group at the MIT Media Lab, the daemon was a "believable agent."

Bots that perform services and act as intermediary between the silicon and the biological are believable agents—the point at which the sets of agents and bots intersect, where a world that arose for reasons of fun and games meets a world dead serious about simplifying the interaction between human and information overload.

"Something powerful happens with personification," says Maes, a Belgian-born researcher whose sterling record in the fields of artificial intelligence and artificial life has made her a premier authority on software agents. "People engage with videogames. They identify with the characters. If we can capture some of that engagement in useful software—software designed for more than pure entertainment—this would be really, really powerful."

CYBORG DREAMS

"We encounter the deep questions of design when we recognize that in designing tools we are designing ways of being," writes Terry Winograd, the onetime MIT artificial intelligence researcher who later renounced many of the core precepts of the AI community. "The use of technology . . . leads to fundamental changes in what we do, and ultimately in what it is to be human."

The use of bots as tools—their merger with agent technology into the new category "believable agents"—requires a closer look at what lies behind the human tendency to anthropomorphize. On the one hand, endowing software programs with character and personality does make them more user-friendly. Thinking of them as fellow beings rubs the rough edges off the stark reality of the human-computer interface. Gamebots, chatterbots, purring kittens, and howling dogs add a little cartoon sparkle to an otherwise drab environment of bits and bytes.

But there's a flip side. The act of personification obscures an important truth about the human relationship with software tools. The dichotomy created between "us" and "them" is false. We're in this together. Software tools—whether we call them bots or agents, whether they are believable characters or not—are extensions of ourselves, prostheses that we use to manipulate objects outside our flesh-and-blood day-to-day life.

We are one with our tools—an observation that sheds new light on the messy issue of defining exactly what is a bot. In answer to a question that attempted to pin down the technical difference between an IRC bot and an IRC script, one IRC hacker, John Leth-Nissen, dismissed the query as functionally irrelevant. "We can argue all day over whether a script should really be considered a bot," he said, "but what we can't deny is that a human using a script or a bot is a cyborg."

A cyborg. The term reeks of science fiction fantasy. But it is a powerful metaphor through which to view the human-computer interface. Mailbots, search robots, warbots, cancel-bots—all bots that have a purpose are really extensions of the human using them, not autonomous beings in their own right. Agents aren't being invented solely because computer scientists and corporate researchers are lonely. Agents are being designed to solve problems that humans can't cope with: to move those molecules of gas from one side of the box to another, to go where no human has gone before.

The logical leap from bot surrogates and agent helpers to full-blown cyborgs becomes all the more obvious when you listen to some of the leading agent provocateurs talk about their ultimate dreams. Pattie Maes, the advocate of believable agents, offers the best example. At conferences and in interviews during the summer of 1996, Maes punctuated every spiel she delivered on agents with the latest buzzword to come out of the MIT Media Lab idea factory—"wearable computing."

Wearable computing is the ultimate incarnation of dreams of agent empowerment, a human-computer synthesis that transcends all the limitations inherent in the human-computer interface. The interface doesn't have to stop at the computer screen. At MIT, plans are currently being made to wire bot and agent helpers right into the body. Graduate students wander the halls with bizarre Borg-like contraptions strapped to their heads, liquid display readout screens suspended over one eye. In addition to video cameras recording everything they see and hear, they are equipped with keyboards strapped around their waists, battery packs and processing units located in strategically comfortable positions across their bodies. There's even talk about getting rid of all the cumbersome cord connectors, about using the body itself as a low-voltage transmission conduit for data and a source of electric power.

Why is Maes so fixated on wearable computing? Because, she says, she is frustrated at her own inadequacies. It seems she is bad with names. She hates seeing people she recognizes and not being able to remember who they are. She longs for the day when her onboard computer will be tracking every person around her. The idea of having a computer equipped with vision recognition analyze each face that approaches, compare it with a database, and then pop the correct name onto her discreetly head-mounted miniterminal is seductive.

Maes is frustrated with her biological limitations. She can't wait for a legion of bot and agent helpers to be at her beck and call, advising her on every contingency and dealing with her email overload, her scheduling conflicts, her forgetfulness and time-management inefficiencies. And her frustration, on a broad scale, is one of the major motivational forces behind bot development. Agentbots will make us into superior beings, smarter, more efficient, faster, stronger. Listen to one MIT graduate student describe the advantages of better living through wearable computing: "We are on the edge of the next stage of human development: the combination of man and machine into an organism more powerful than either."

An organism more powerful than either human or machine. Is that the bot way of being? Is that where the headlong rush toward an ever more anthropomorphized human-computer interface is leading? Is it a quest for superiority? Certainly, that goal jibes with the dreams of a sizable segment of the artificial intelligence community. Human-level intelligence is not their Holy Grail but rather the surpassing of humanity's limitations, the creation of an intelligence greater than what mere mortals are capable of. Agent enthusiasts and AI workers alike want machines and software that are unsusceptible to human frailties, that transcend the restrictions of flesh.

Cybercultural critic Mark Dery calls the passion for cyborg superiority "a misguided hope that we will be born again as 'bionic angels' . . . a deadly misreading of the myth of Icarus [that] pins our future to wings of wax and feathers."

Dery's finger is on the pulse of our bot desires. The original Czech meaning of "robot"—"forced labor"—reveals the theme of superiority inherent in the concept of automaton servants from the very beginning. Bots, whether employed as interface, intermediary, flat-out tools, or chatterbot companions, can't escape their human-made nature, can't escape the dominance-submission relationship essential to the bot way of being. Bots are servants and slaves. And there's something unsavory about that. They get no compensation for their service; they have no true freedom of action, no options. They are the method by which humans wield power in the digital realm.

Agent enthusiasts think it's neat to have digitally indentured servants. MIT's Nicholas Negroponte wants his digital butler badly. But what kind of illusions are we perpetuating when we dream of marshaling armies of bot soldiers to do our bidding? As is so often pointed out, power corrupts. Like Icarus, are we heading for a fall?

WAR

"THE POTENTIAL FOR ABUSE HERE SEEMS LARGE"

This is the robot construction laboratory. There's a wooden cabinet over in the corner, and an exit to the west. Also, there's a trash bin off to the side.

fur says	"robot ethics, an interesting can of worms . . ."
Majik says	"I've heard there is a robot that goes around killing people. This could be annoying."
fur says	"Personally, I don't have any problems with what anyone (or any robot) does, as long as it's done in moderation."
Majik says	"real world ethics, though. Could a robot hog the entire system and crash it? The potential for abuse here seems large."
Fuzzy says	"An interesting question is how 'secure' robots have to be. Is it ethical to fool other people's Robots, and what constitutes a good puzzle that a robot builder must handle, and what constitutes harassment?"

Majik says "a world of robots. I wonder what that would turn out
to be like?"

The hubbub in the laboratory adjoining TinyMUD's Recreation
Room on the night of December 9, 1989, marks a turning
point for bot-human relations. Over the course of the evening,
some twenty-five beings, human and not quite, crowd into the
virtual space, rubbing their packet-switched electrons together
in convivial glee. They have come for a historic reason—the
First TinyMUD Robotics Conference.

Potential eavesdroppers are likely to be intrigued. The
group boasts a set of typical *handles,* MUD nicknames that,
once again, owe much to the etymology of *Lord of the Rings:*
Hobbit and Wizard, Prothan and Froboz, Majik and MisterX.
But the multitude of online hobbits and warlocks isn't what
makes the laboratory the place to be in cyberspace this particular evening. The bizarre snippets of technogeek jargon and
existential philosophizing emanating from the laboratory are
the real draw. How does a comparison of the relative merits of
the C++ and Lisp programming languages intersect with concerns about ethics and responsibility? What do "forking
processes" and "porting socket code" have to do with questions of robot hostility and sneakiness?

Everything. In the robot world, separating questions of
internal combustion from behavioral psychology is neither
possible nor desirable. The "mental" makeup of a bot is a Skin-
nerian wet dream—for every action there is a line of code, or
an environmental catalyst, that can be held accountable. For
the group gathered together this winter evening, untangling
the implications of robot cause and effect is indeed the main
item on the agenda.

Described by one attendee as "a big get-together for everyone interested in robots," the conference has an open-door

policy. In one corner of the room sits Fiona, a robot secretary (and self-described ace stenographer), scribbling away industriously with pen on pad. Gloria, one of Michael Mauldin's early Maas-Neotek chatterbot prototypes, also attends.

But the bots don't contribute much to the discussion. It is the humans, the bot authors and bot owners, who do the talking, who take the first exploratory stabs at understanding some of the deeper issues of bot existence. What indeed would a world of robots be like? What truly are robot ethics?

< >

For bot aficionados whose images of the future are shaped by science fiction dreams, the phrase "robot ethics" evokes powerful visions of human-machine conflicts played out across the panorama of future history. Isaac Asimov's famous robots had their Three Laws traced into the positronic structure of their brains—their ethics were immutable and required clear subordination to the human race:

1. A robot may not injure a human being, or, through inaction, allow a human being to come to harm.
2. A robot must obey the orders given it by human beings except where such orders would conflict with the First Law.
3. A robot must protect its own existence as long as such protection does not conflict with the First or Second Law.

Handbook of Robotics, 56th edition, A.D. 2058, as quoted in Isaac Asimov's *I, Robot*

Robots aren't always so constrained. In the dystopian cyberpunk future of science fiction author William Gibson, the scheming "AIs" that inhabit the interstices of the information matrix transcend human ethics—they are accountable only to themselves. In innumerable alternate realities, the robotic struggle with the fuzzy and imperfect parameters of ethical behavior is a recurring plot fixation, an essential trope of the sci-fi genre.

A capacity for ethical behavior implies the ability to discern between right and wrong. But venturing onto such unsure ground presents robot designers with problems that dwarf the obstacles already daunting the artificial intelligence industry. AI workers have yet to solve nuts-and-bolts AI tasks such as vision recognition and natural language processing or to come close to instilling machines with common sense or the ability to learn. In light of those failures, how can robot designers even begin to discuss the implementation of a decision-making process for ethical questions?

"Robot ethics" is an inadequate formulation of the problem. As "fur" observed at the Robotics Conference, "A robot can only be as sneaky as the person who programmed it." Until robots really do learn to think for themselves, their every action will be a reflection, purposeful or accidental, of their programmer's intent. A more appropriate framing of the issue might be "the ethics of robot design" or "programmer ethics." It is not yet time to blame bots for the sins they commit. As the submissive half of the cyborg human-computer interface, bots are not responsible for their misdeeds. Humans are.

BOTS GONE BAD

At Point MOOt, Allan Alford and his fellow wizards discovered early on that their bots tended toward perverse and unexpected behavior. Long before the Barney plague racked the plains of virtual West Texas with its saccharine miasma, equally strange phenomena could be observed over on the wrong side of the Point MOOt tracks, in the bad part of town. In those early days, you did not want to hang around when a bum bot chanced upon a hooker bot. Things got ugly fast.

That bums and hookers might occasionally bump into each other comes as no surprise. One might even imagine that the seedier neighborhoods of a city based on the credo of "reality modeling" would welcome clashes between such stalwart representatives of the underclass. But in Point MOOt, every bum-hooker tête-à-tête invited disaster.

Bum and hooker bots possessed limited English-speaking skills. When in the same room as other Point MOOt citizens, their programming required them to ask a simple, straightforward question and then wait for a specific answer. Inexorably patient in typical robot fashion, they never tired of requesting spare change or making lewd propositions. At least, not until they received a satisfactory answer, in language they could understand.

So a hooker bot saunters over to a bum bot and makes her pitch. The bum bot's natural response is to counter with its own plea for MOOlah. But the hooker bot's pea-size brain doesn't know how to parse that kind of answer. A yes or no reply it can handle, but not a demand for a handout. Confused, the hooker bot falls back into default mode and repeats the original solicitation. Undaunted, the bum bot, programmed to keep pestering whomever it encounters until it gets some money, panhandles again.

The cycle, once started, will not end. Bum and hooker face off on the mean streets of Point MOOt, locked in a verbal loop, trapped in a vortex of infinite recursion. And since they are both bots, their quick-draw reaction time seems instantaneous to a human. Only the smartest of chatterbots know enough to dawdle.

"We called it a hideous spam-convulsion loop," says Allan Alford. "Two sentences would scroll by, over and over again, at ninety miles an hour."

Robot insanity.

Infinite recursion is the most common bot neural disorder. Its consequences, depending on the situation, can be devastating to system or network computing resources. And it is by no means confined to MUDbots. Web robots get trapped in black holes of infinite recursion as they navigate the intricate topography of Web servers. IRC warbots are designed to use infinite recursion as a weapon—a technique for disrupting IRC channels or overloading specific IRC servers.

Mailbots are especially good at infinite recursion. Just before Christmas 1995, a vacationing subscriber to an email discussion group devoted, ironically, to such issues as how to avoid the occurrence of infinite recursion in robots, instructed his mailbot to answer each incoming piece of mail with a return message stating that he would be unavailable for the following two weeks—so please don't send him any more mail. Unfortunately, as a member of the email group, he received all the group mail that was sent to every subscriber, including his own mailbot messages. The vacation mailbot failed to recognize its own alerts and responded to them with the same old vacation notice, again. And then to its own responses to its own responses. And so on. In a matter of hours, the discussion group exploded out of control, a victim of a runaway robot. Only the group's moderator had the power to do anything; the rest just had to sit and watch as their mailboxes overflowed with email garbage.

Infinite recursion is, of course, a bug. A glitch, a programming failure, a huge bot no-no. Although programmers tend to view bot bugs as errors of omission rather than commission, those bugs can have a debilitating effect on the look and feel of an online environment. And they are not rarities. They are rampant. Bots are just beginning to crawl out of the primordial digital ooze, and all the kinks haven't been worked out of their genetic code. In the real world, they are imperfect beings.

"Every new prototype we developed would screw up," says Alford.

A bug does not even have to be technical in nature to be destructive. Any failure, technical or conceptual, to anticipate how a bot might interact with its environment can be considered a programming error with ethical implications. Not only were the hooker bots of Point MOOt, for example, prone to infinite recursion psychotic breakdowns, but the earliest models also consistently alienated female citizens of Point MOOt by their failure to be discriminating.

"If you've got a female hooker walking up to a female player [and] saying, 'I want to suck your dick,'" says Alford, "it doesn't gel. It shatters the illusion."

At your run-of-the-mill MOO, gender trouble might be good for a laugh or two, but at Point MOOt, where the advising professor was a transsexual postmodern critic specializing in identity issues, gender was guaranteed to be a sensitive subject. A failure to select anatomically correct solicitation options constituted a bad bug.

But so what? Bugs are easy to squash. No big deal. Constant tweaking is an integral part of the programmer's life. If the program misbehaves, then fix it. Ethical lapse? Hardly. An implementation issue—that is all.

< >

But what if the program behaves exactly as intended and still causes problems? What if the program is designed for the very purpose of wreaking havoc? What if Cthulu is on the rampage?

Cthulu, Point MOOt's darkest secret, a twenty-five-foot-high grayish green monster with hideous tentacles curling out of his head, occasionally erupted on the town of Point MOOt like a nightmare ripped straight from the pages of H. P. Lovecraft.

"He was covered with slime, had evil death eyes—he could kill you with his gaze," says Alford. "He lived at the bottom of a lake at the edge of town, and randomly, once every great while, Cthulu would emerge from the lake, pick a random direction, and stomp through town, causing earthquakes and crushing dwellings.

"Cthulu was this unstoppable force, a weird sci-fi big brute of a bot," says Alford. "Then one of the wizards went too far and gave Cthulu a big net sack that could capture citizens."

Alford remembers logging on one day and discovering that Cthulu had snatched up ten residents of Point MOOt and stuffed them in his sack.

"'Help me, help me,' they cried," says Alford, who at first found their plight amusingly ludicrous. But the joke soon stopped being funny. As Alford began to inspect Cthulu's code, looking for the trick that would let the citizens out of the bag, he too fell victim, jumbled into Cthulu's sack with the rest of the Point MOOt hoi polloi. The only way to get free was to log off entirely and then log back on. But each time Alford logged back on, he got nailed again.

"Even though I had wizard access," says Alford, "the code was so deep and so thick that I couldn't figure out how to stop him from nabbing us."

Finally, Alford wrote a "nasty letter" to Cthulu's author, ordering him to make some changes in Cthulu's programming.

Sure, Point MOOt had dreams of democratic self-government, but you messed with the archwizard at your peril.

BOT AND ANTI-BOT

Mischief informed Cthulu's behavior, not malice. But not all MUDbot misbehavior can be dismissed as cheap thrills. Some people employed bots to get the upper hand on their fellow MUDders. They were motivated not by mischief but by blood-thirsty competition or greed.

Although TinyMUD was primarily social in nature, like Point MOOt it had an economy of sorts. You needed money to be able to build your own rooms or objects. But TinyMUD did not require that you put your nose to the grindstone and start slaving away, nine to five, as did Point MOOt. Instead, pennies lay scattered around the MUD for people to find.

TinyMUDders soon discovered they could easily write a bot that would explore the MUD and grab all the pennies— the ultimate get-rich-quick scheme. To some TinyMUDders, penny-snatching bots ran counter to the spirit of TinyMUD. What was the point of setting up elaborate rituals if someone just wrote a bot and trashed a way right through them?

The bot-as-surrogate (or bot-as-aide-de-camp) ploy raised hackles in purely competitive game-based MUDs. In a compet-itive MUD, one of the goals is to gain points and levels in order to rise up in the MUD hierarchy, to go from lowly peasant to awesome mage. Such games require the player to pass numer-ous tests, solve puzzles, or engage in frequent combat. If a player has a few helper bots that can move fast—killing any-thing that moves, grabbing treasure, and gulping down food— then that player gains what other players see as an unfair ad-vantage.

MUDs became divided into bot and anti-bot factions. Roo

Goldsmith, an inveterate MUDder, summed up the anti-bot position for me in an email message:

> Using a bot is unfair and no one should ever do it. So I spend 4 hours a day playing on a MUD trying to reach the highest level and be the biggest baddest person on the MUD. I log my hours, lose my sleep and generally give everything I've got and then some to being the best.
>
> On the same MUD a "pro-bot" player puts in his 4 hours a day doing the same thing, killing the same stuff as I do, etc. We're friendly competitors in the quest to be top dog. Difference is the "pro-bot" player also leaves his character on-line the other 20 hours a day as a bot. So what happens? I'll tell you what happens, I get left in the dust as his bot does all the work for him.
>
> On MUDs where you have to be creative and a bit lucky to gain levels, bots are not terribly useful. [But if] the MUD you are playing on is a simple kill-the-monsters-and-everything-you-kill-will-eventually-get-you-to-a-higher-level MUD, bots can be a great advantage. Turning a bot loose on easy enemies will gain you levels, especially if you can run your bot all the time you are off line.
>
> Now there is a third category of bots, what I like to call the "cyborg." These are players who use a telnet client to automate the majority of their character's actions. For example a simple cyborg would eat and drink automatically, the player never having to worry about anything more than providing food. More complex cyborgs will change armor and weapons at the appropriate times and often even cast spells of healing or escape.

"The potential for abuse here seems large," said the TinyMUD bot specialist at the First TinyMUD Robotics Conference. No kidding. When the goal of the game is to encourage clever use of spells or deft swordplay, it seems obviously unfair, if not patently unethical, to use bots for backup or to program your own private cyborg extensions. And what's to stop you from siccing your bot on another player, instead of against computer-generated monsters? Who decides what is fair and what isn't? Who decides what is merely a tool and what is a weapon?

Bot malfeasance is not limited to MUDs. Everywhere that there are bots, there are bots gone bad. Everywhere in the on-line universe, bots are being used for self-protection, for warfare, for financial gain, and for outright evil. And, if anything, as bots have evolved—from MUDS through IRC and Usenet and into the World Wide Web—so has bot mayhem. Even as bots become ever easier to obtain and ever simpler to manipulate, their online environment is becoming increasingly complex, commercial, and immune to central control—all factors that encourage or facilitate unethical bot use.

The First TinyMUD Robotics Conference resolved very little, and the records do not reveal any second such gathering. But one could argue that the whole course of bot evolution on the Net throughout the 1990s has been a continuation of the themes raised that winter evening. Whether discussing game-bots or intelligent agents, the same questions come up again and again. Should a robot be designed to be superior to a human being? Should robots be able to "kill" humans? Should they be allowed to tell lies, take unilateral action, or exchange information with other robots?

At Point MOOt, the point *was* moot. Allan Alford and the other wizards did not fret about bot warfare. A MUD is a closed society, and wizards enjoy absolute control. Alford reset the parameters for bum and hooker bots so they could not overlap each other's territory as they hustled about. He ordered the hooker bot author to rewrite the hooker-chatterbot interface. No longer would hooker bots automatically assume they were plying their come-hither stares on men and men only. In the aftermath of the great Barney plague, Alford added new commands to the MOOt lexicon, giving citizens vastly increased firepower. Despite his often expressed desire to take a hands-off approach toward Point MOOt's evolution, Alford nonetheless found himself forced to be ultimate arbiter and absolute monarch.

"I would inevitably edit anything I found to be too extreme," says Alford. "We calmed the Hunter Thompson bot down to the point where he hardly ever maced anybody. With Cthulu we killed the net sack concept."

MUDs still have governance problems. Any MUD that succeeds in attracting a significant number of people into an open-ended setting (as opposed to a controlled game setting) can expect political tensions. Some—like MOO pioneer LambdaMOO—are epic in scope. But each MUD is its own relatively tiny kingdom, with rulers who have the power, whether they choose to exercise it or not, to make their intentions the unchallengeable law of the realm.

Not so with the Net or with most other subsets of the Net. In the world of Usenet bulletin boards and IRC chat rooms and hyperlinked Web pages, there is no all-powerful wizard, no chief executive officer who can lay down the bot law and decide what is acceptable and what isn't. On the Net, bots and their botrunners run free.

And in the multifarious channels of the Internet Relay Chat network, just as the likes of Julia took their first baby steps forward in carefully controlled MUD environments, a stunning array of unchained bots began to explore the implications of their freedom. These IRCbots knew no bounds.

ROBOBOT IN THE HOT TUB

Nobody enjoys having a robot barge in right in the middle of sex in the hot tub, even if the so-called hot tub is actually an online chat room and the chances of instant electrocution are small. Passion is passion. And in a text-driven environment where all attention is focused on words scrolling down the monitor screen, the interruptions of an obstreperous bot are distracting, not to mention ardor-stifling.

So no wonder the inhabitants of the Internet Relay Chat channel referred to as #hottub were getting steamed as the new year began in January 1992. Out of nowhere, with no explanation, an automaton with the all-too-appropriate name "Robobot" seemed hell-bent on permanently disrupting one of IRC's most popular gathering places—the hot tub chat room: part pickup joint, part coffeehouse, part X-rated sauna.

Described by one complainant as "a self-replicating, self-inviting, self-nickchanging automaton that rejoins after kick and after kill and is extremely annoying to all hottub users," Robobot seemed designed solely to annoy.

Self-replicating: If you attempted to evict Robobot from the channel by "kicking" or "killing" it—two slices of IRC jargon designating specific techniques for banishing unwanted chat room participants—Robobot always returned, like an online Terminator, relentless, automatic. It would not, could not, die.

Self-inviting: Back in 1992, normal bot etiquette required that a bot be "invited" into any given channel by a current human inhabitant of that channel. Robobot refused to wait for any such solicitation, shouldering its way into #hottub willy-nilly, welcome or not.

Self-nickchanging: One traditional IRC response to persistent intruders is to place a ban on a nickname—IRC's version of a handle. Anyone attempting to enter a chat room under a banned nickname is automatically kicked right back out. But Robobot sneered at futile attempts by puny humans to bar the door. If banned, Robobot, before rejoining the channel, simply changed its nickname, to, say, Robobot41 or Robobot89.

And that wasn't the end of it. Once inside #hottub, Robobot automatically greeted each new arrival to #hottub with a canned welcome message and bid him or her adieu with an equally predetermined farewell. In a crowded chat room like #hottub, the constant stream of mindless hellos and good-byes promulgated by an indiscriminate autogreet function clogged up valuable monitor screen real estate. Robobot also periodically left the channel for no perceivable reason and then rejoined, a habit that also generated text messages bound to disrupt ongoing conversations in the hot tub.

But the fun really started when anyone was so bold as to attempt to kick Robobot out of the channel. Robobot's self-replication abilities incorporated a nasty twist. A member of the IRCbot subspecies *clonebot,* Robobot could spawn multiple identical Robobot processes. Robobot did not content itself with merely rejoining the channel or regenerating itself after being killed. After receiving a #hottub eviction notice, but just before being (temporarily) squelched, Robobot cheerily exclaimed, "Gotta go!" Then, seconds later, it rejoined the chat room, along with a bot compadre, a Robobot doppelganger. Robobot number one announced, "I'm deeply hurt . . . so I'm bringing a friend." Robobot number two then chimed in: "Hi!"

Expel both the new Robobots and the process repeated itself, after another split second. Only this time four robo-brethren marched into the chat room. Such behavior, combined with a defensive auto-kick program that automatically expelled each new Robobot incarnation, meant that an ever increasing horde of Robobot clones flooded the channel, bringing to an effective halt any chance for hot tub conversation or attempted seduction.

The hottub channel already had its own bot, Hottubsrv. Hottubsrv performed a number of help menu functions and other specific #hottub services. For instance, if you typed the word "towel," Hottubsrv "handed" you a hot towel. In contrast to the anonymously operated Robobot, Hottubsrv, if inspected, carried with it the email address of its operator, a Princeton student named Peter Hellmonds, who administered the local IRC server. If Hottubsrv was kicked off the channel for some reason, it patiently waited until expressly invited back in before rejoining. Hottubsrv, to most #hottub regulars, behaved quite nicely.

But Robobot, the bane of the #hottub chat room, behaved like a churl.

< >

By the beginning of 1992, bots were old hat on IRC. Ever since Jarkko Oikarinen included the source code for a generic IRC automaton in the original release of IRC, IRCers and bot development had gone hand in hand.

The first IRC explorers knew their way around a computer. Setting up an IRC server isn't a trivial undertaking and usually requires access to a permanent Internet connection. In the late eighties, the kind of person who enjoyed that type of access tended to be either a student of computer science or a professional in the field. And even if you weren't the kind of systems administrator who might have originally set up an IRC server, getting the most out of the primitive UNIX-style commands necessary to use IRC required a basic familiarity with programming concepts. To the early IRC adopter, Oikarinen's source code offering was a gesture akin to pushing digital candy on hacker babies. No further encouragement necessary.

IRC's basic architecture, which facilitates script writing, made it easy. A script is a simple program that automates a set of procedures normally done manually. Experimenting with scripts is an obsession with most programmers, who are maniacs about simplifying and streamlining their every interaction with the computing environment. The earliest IRC hackers fit the profile, psychologically identical to Fernando Corbato's team of MIT researchers nearly three decades earlier. IRC, where the steps from a script to a bot are short, soon became awash with daemonbots, programs permanently logged on to IRC, accessible from any channel, and willing and eager to serve (thus the surname "srv" or "serv").

One early bot, written by a Norwegian, helped IRC users solve identity questions. If given a person's name, the bot searched an internal database and responded with the name of the country that the name most likely came from, along with the probable gender of the name—theoretically a useful tool for attempted pickup dating maneuvers, which could be confusing in an international and virtual setting. As the author of the bot noted, the name "Kari" is female in Norway but male in Finland.

Another tool attempted to solve the problem of nickname stealing. IRC users choose a nickname when they log on to IRC, and some people prefer to use the same nickname every time they log on. Unfortunately, a nickname stays yours only as long as you are actively logged on. As soon as you break your connection, someone else can grab your nickname and then it won't be available to you when you next log on. The bot NickServ monitored nickname registrations. If you chose a nickname that belonged to someone else, it notified you that the nickname was off-limits. In some cases it might even try to block an unregistered nickname grabber from connecting to the network.

Like MUDs, IRC chat rooms are bot-friendly; they provide all the safety of a Turing Test without any of the suspicion. Some of the first IRCbots were functionally identical to the early place-keeping bots on EssexMUD: simple programs that logged on to an IRC server, "joined" a specific channel, and then periodically emitted brief statements in imitation of humans. Crude natural language tricks soon became popular, just as had been true in MUDs. Primitive bots called *ircskels* joined channels on request and then constructed inane commentary, using randomly chosen words from ongoing conversations.

"It seemed to provide great amusement to put a bunch of these on the channels and let them speak to each other," remembered one early IRC pioneer. Part of the amusement lay in setting up opportunities for infinite recursion, although that usually lost its attraction after one or two disastrous instances.

By the height of the bot era, IRC boasted bots that could spellcheck channel conversations, spy on chats and report back to the absentee bot owners, or simply keep logs of everything that happened in a given channel. Virtual bartender bots mixed imaginary drinks and made idle chitchat. Bots played *Scrabble, Jeopardy, Boggle,* and other games. Help-menu bots offered automated access to information on topics: they could tell you how to prevent a rape, steer you to the latest *warez*—pirated illegal software—or even deliver that software to you automatically, no questions asked. Bots could link channels together on different IRC networks or convert Fahrenheit temperatures into Centigrade and back. In channels frequented by devout Christians, biblebots responded to keywords with chapter and verse from the King James Bible—in multiple translations.

Bots could be purely silly—at one time #advice featured a bot called Doc-Ruth that responded to any mention of the word "pickup" with a sample pickup line. But they could also be simply utilitarian—some system administrators ran bots designed solely to keep a watch out for other bots.

In the annals of bot evolution, IRC in the mid-nineties will be remembered by paleobotologists as a modern-day equivalent to the Cambrian explosion—a relatively short period of time 570 million years ago that spawned more new species of life than ever before or since. Until recently, IRC ranked as the most fertile environment for bot creation in the online universe. The distributed nature of its thousands of channels encouraged experimentation by hundreds of bot authors. The shared architecture of all those channels meant that those same bot authors could trade tips back and forth, copy the best parts of other bots for their own use, or simply import useful bots as they saw fit—activities not possible in the fractured world of MUDs.

True, just as with species in the early days of the Paleozoic era, most IRCbots fast became extinct. The steady advance and upgrading of the IRC protocol required that bots be constantly rewritten, and if the original creator wasn't interested, sometimes the bot was out of luck. (After one such upgrade, Greg Lindahl never bothered to rewrite the *Hunt the Wumpus*–moderating gm bot.) In less than a decade, IRC had accumulated a bot fossil record flush with obscure specimens, doomed to eternal footnote status in the bot chronicles.

ANARCHY HAS ITS FLAWS

Among a swarm of service-oriented bots, gamebots, and assorted mindless *lamebots*, the havoc wreaker Robobot stood out. Clearly, its only purpose was to annoy, and eventually one regular #hottub habitué nicknamed Scorpio revealed that the Robobot had indeed been written by a disgruntled IRC user angry at a perceived slight from Scorpio in another channel. But Robobot wasn't unique. It was far from the first malicious IRCbot. Long before Robobot busted into #hottub, nefarious-minded bots had popped up here and there. One IRC veteran remembered a bot that sat in a channel waiting to "kill" particular users should they ever show up. Another early IRC user programmed a bot to grab and hold nicknames of users he didn't like, just to spite them. A bot known as Revenger monitored channels for every instance when another user was "killed"—and then "killed" the original killer.

Robobot is significant not for any special technical wizardry or particular concentration of malevolence but as a marker for a changing IRC and, by extension, a changing Net. Robobot's assault on #hottub came at a time when IRC's skyrocketing growth rate had begun to place unforeseen stresses on the network. The advent of Robobot signaled IRC's transition from a small, insular community where everyone knew everyone else to a much larger, more diverse population full of cliques and strangers, generation gaps and power conflicts.

In early 1988, IRC servers existed only on three machines in southern Finland. But IRC's growth curve soon pointed straight up. By mid-1992, five hundred people could be found using IRC at any given time. By the end of 1994, the number had risen to five thousand. A year later, fifteen thousand. By 1996, twenty to thirty thousand people simultaneously chatted online in thousands of different chat rooms via IRC servers scattered across the globe.

All this growth occurred without any semblance of practical organization, exactly as Jarkko Oikarinen had intended. Oikarinen accompanied the release of the IRC source code with his own withdrawal from active modification of the IRC protocol. IRC users, if they wanted to improve the network, had to do it themselves. The inhabitants of IRC weren't just masters of their domain; they were the architects, craftspeople, and handymen entrusted with building, maintaining, and fixing it.

IRC has no government per se—no constitution, no set of laws, no police force. On IRC the closest thing to an organized decision-making structure is a set of email discussion groups (also known as mailing lists) for policy debate, announcements of technical updates, and gossip. But no one is bound by the consensus opinion of these groups, and, indeed, consensus is rare.

Some rules of bot etiquette evolved and became codified. Bots should be identified as such and should bear a suffix identifying their true nature, such as "bot" or "serv." Bots should be killable—an invulnerable bot constituted a serious failure to respect bot etiquette. Bots should not speak unless spoken to. But these rules were obeyed as much in the breach as in practice.

Peter Hellmonds, the Princeton student who ran the Hottubsrv bot and administered that New Jersey IRC server, complained about Robobot in a post to a mailing list frequented by IRC operators—the administrators of IRC servers. He also took the dramatic step of sending an email message alerting a system administrator at the University of Wyoming, where the Robobot process had originated, of Robobot's crimes. The system administrator, who knew nothing about IRC, immediately took the drastic action of shutting down the local IRC server, thus cutting off about fifty regular IRCers from their connection to the IRC network.

Hellmonds' peers reacted harshly to his complaint. One correspondent demanded that Hellmonds "quit being such a fascist and trying to 'own' a channel." She added, "Grow up, get a life, and find something better to do with your time than whine and moan about your precious channel being invaded." Other commentators offered technical advice on how to neutralize Robobot without resorting to calls to higher authority.

Contacting the University of Wyoming system administrator pressed exactly the wrong button. IRCers treasure their freedom. Like so many early Internet pioneers, they cherish the anarchy of an open-ended network without centrally imposed laws, autocratic dictatorships, or tyrannical majority-rules policies. By seeking outside help for an internal matter, Hellmonds, or so the ICRers believed, had taken a step that could doom all of IRC civilization. Far better to ignore Robobot's depredations than to give the evil powers of the State an excuse to clamp down.

Robobot disrupted #hottub, complained the #hottub denizens. So what, responded the IRC old guard. Go somewhere else. IRC had no limits. If someone or something disrupts one channel, anyone who cares to is perfectly free to start a new channel—#hottub2.

That is the beauty of IRC—anyone can start a channel, on any topic, for any reason, at any time. If you want to talk about vampires or obsolete computer operating systems or dynastic politics in Bhutan, no problem. Just fire up the IRC client software, connect to an IRC server, and create a new channel. All you need do is type the command "join channel #x." If the channel already exists, you immediately join it, providing you haven't been banned from the channel for previous indiscretions or the channel isn't designated "invite only." But if the channel does not exist, then it is instantly created, and the first inhabitant of it is you, the creator.

Grow your own channels. It isn't just the motto of IRC—it is a fundamental principle of the Net, a core value of the libertarian ideology that once reigned supreme in cyberspace. IRC users, like those who participate in Usenet or flock to the Web, flourish without central authority and scorn top-down solutions. From its roots in a Defense Department project to link major research institutions, the Internet has grown in an unscripted and unpredictable fashion, with rules being made up on the fly by its own users. What rules there are—the bot etiquette of IRC, the elaborate protocol for the creation of new Usenet newsgroups, or even the adoption of HTML as the language of the Web—occur almost of their own accord, through processes of bottom-up evolution rather than as laid down by fiat from above.

Or so went the conventional wisdom. To an outside observer, conclusions could vary. The early Net-based communities enjoyed the benefits of a homogeneous citizenry: mostly white, male, well educated, and working in the high-tech sector of the economy. These netizens shared the same values, and in their relatively small communities (as compared with the outside world) they could work out problems in an informal, ad hoc manner.

But for how long? That was the question Robobot unwittingly asked as IRC and every other Net-connected society began to explode in size. As the nineties progressed, the same generalizations about shared values no longer applied. The Net's userbase was getting both older and younger, more international and more diverse.

And more prone to mischief. Robobot was just the leading edge. An army of warbots lurked in IRC's future—floodbots, clonebots, annoybots, collidebots, crashbots, and more. Take-over wars, botrunning gangs, and pitched battles raging across scores of channels loomed on the near horizon. Before Robobot appeared in 1992, IRC's small size made each incident of bad bot behavior a minor cause célèbre for the whole IRC community. But by the end of the year, a bad-bot epidemic had begun to sweep IRC. And there seemed to be only one good solution to the dilemma posed by proliferating bot pests. More bots!

EGGDROP

/JOIN #gaysex

Habanero: >Hi B-Town! I'm Habanero, an eggdrop bot.<
Habanero: >All commands are done via /MSG. For the complete list, type:<
Habanero: >/msg Habanero help<
Habanero: >Cya!<

B-Town: /msg Habanero help

Habanero: >MSG commands for Habanero:<
Habanero: >EMAIL INFO WHO IDENT GETBOYFRIEND GETGIRLFRIEND<
Habanero: >For help on a command, type /MSG Habanero HELP <command><

Robey Pointer well remembers that fall day in 1991 when for the first time ever, five hundred IRC users chatted online at the same time. "We had a little online party," says Pointer, smiling. "We just crossed five hundred!"

At the Mexican restaurant deep in the heart of Silicon Valley where he is having lunch, Pointer seems both amused and proud at the memory. Amused, because in the great scheme of things, five hundred people online at the same time is not quite a world-shaking event. But proud, as well, because the moment served as a welcome benchmark—proof that Pointer's then newfound community exhibited vigor.

At the time, Pointer was a freshman at college in South Carolina. A self-taught hacker and a regular patron of local bulletin boards while still in high school, Pointer had discovered IRC only a few months earlier. But it was love at first sight. IRC offered what Pointer needed—companionship and support.

Like many other newcomers to IRC, Pointer gravitated first to #hottub, one of the most popular general-interest social channels on the network. But it wasn't long before he moved on to a new channel—#glbf. The initials "glbf" stand for "gays, lesbians, bisexuals, and friends." As it turned out, the fall of 1991 wasn't notable merely for IRC's landmark progress; it was also a breakthrough period for Pointer. That winter, Pointer says, he came out to himself as a gay man.

In South Carolina in 1991, for the eighteen-year-old Pointer to acknowledge his own sexual orientation took considerable courage. South Carolina is not renowned for its teenage gay support groups. IRC, however, is an infinite network of support groups for every possible orientation, ideology, fetish, or fixation. In chat rooms like #glbf, Pointer found friends who could give him the emotional safety net he lacked in his face-to-face real-world life. Like countless others whose circumstances or predilections left them isolated in the physical communities in which they happened to reside, Pointer thrilled at the discovery of one of the key selling points of on-line life—its facility at creating "communities of interest." The effortlessness with which the Internet transcends physical boundaries of location or time allows interested parties to get together to discuss any topic at their convenience. Even in a big city like, say, New York, it might be difficult to meet a quorum if the only topic on the table is something as obscure as the operating idiosyncrasies of antique Frigidaire refrigerators. On the Net, it's a breeze. To take just gay sexuality as an example, although in 1992 only #glbf offered succor for those who felt so inclined, by 1996 scores of gay-related channels flourished—everything from #gaytruckers to #gaysanjose.

For those whose chosen topic is not just obscure but also the subject of active disapproval in the "real" world, the advantages of the Net are even greater—and often overlooked by those who bash online existence for its supposed disassociation with the physical world. For Pointer, IRC became a place where he could talk about his own sexuality without fear of physical abuse. It became a place of priceless value. And he would do anything to defend it.

At first he didn't need to do anything at all. In late 1991 and early 1992, the specter of the Robobot notwithstanding, IRC was still small enough and close-knit enough to have the flavor of a small town where everyone knows everyone else's secrets. The atmosphere tended to be friendly.

"Back then, bots weren't anything like they are now," says Pointer. "You could join a channel and someone might have a little script set up so that anytime someone joined, it would say hello and then your name. Like, 'Hello Wizard!' or 'Hello Robey!'"

But the times soon changed. Before long, says Pointer, "people started saying, 'Whenever this person joins a channel, I want to op them because they're my friend.' And they started putting that function in bots. And for a while, that was mostly what bots did: keep a list of people that were supposed to be op'ed and maybe keep a list of people they didn't want to be op'ed. And that was it."

Op'ed?

Ops—short for "operating privileges"—are the crux of IRC politics. More virtual blood, sweat, and tears have been shed over their gain or loss than over any other single element of IRC life.

The original creator of an IRC channel gets channel ops. They include such powers as the right to determine whether or not the channel is open to all comers and, of those allowed in, who is permitted to speak. Any IRC user with channel ops can give any other user ops or can kick users out of a channel. But there is one big catch. If a user logs off, his or her channel ops are extinguished, and this is true even for the original creator of a channel. If someone else logs on and re-creates a popular channel (channels exist only as long as someone is actually in them—as soon as they are empty they disappear), then that person has ops.

The possession of channel ops, or the lack thereof, separates occupants of a channel into two virtual classes—those with power and those without. It is a situation ripe for encouraging the worst kind of cliquish petty politics, a setup that guarantees that the most popular channels are awash in a constant flow of op bestowals and op conflicts. The fight to keep or grab ops, more than any other single factor, ensured that bots would become vital players in IRC life.

First came the bots that kept lists of who should be automatically op'ed or banned. The next step was to ensure that these bots, referred to as channel protection bots, remained online even when their owners were off, thus preserving the ops status quo as per the desires of the channel creator. Channel protection bots became an IRC cottage industry, with names ranging from the silly (the BalooBear bot, the VladBot) to the suspicious (the Hackbot, the Combot).

Cliquish rivalry wasn't the only reason that IRC users clamored for the source code to state-of-the-art channel protection bots. As Robey Pointer soon discovered, a rising tide of abusive behavior coincided with the steady increase in popularity of IRC. That very quality of anything-goes openness that made IRC so attractive to Pointer also encouraged a pervasive disregard for civilized niceties like good manners or tolerance of nontraditional lifestyles. The absence of any real accountability, a comprehensive access to anonymity, and a widespread attitude that rude behavior is acceptable online even when it would be considered in gross bad taste offline contributed to a steadily degenerating quality of discourse.

Any channel devoted to a controversial subject—abortion, for example, or the teachings of Rush Limbaugh—could be counted on to receive unwanted attention. Any channel that attracted regular crowds—channels that hosted online games or that were designated as hangouts for self-described hackers or "warez doodz"—seldom experienced periods of calm. But nowhere was the problem more severe than in precisely those channels that Pointer felt most comfortable in. Pointer remembers a steady stream of drive-by–style attacks in which troublemakers cruised through his favorite chat rooms spewing homophobic slurs.

By the end of 1993, two channels Pointer regularly hung out in, #gaysex and #gayteen, were being attacked and harassed daily. And the protective measures in place were proving inadequate.

The #gaysex channel had a bot named Cevin (named after a singer in an industrial hard-core band). Part channel protection bot, part helpbot, Cevin handled numerous chores. If you sent Cevin a message asking for help, he'd spew out three paragraphs of advice and rules for how best to behave in the #gaysex channel. He also had a list of banned troublemakers and a list of automatically op'ed friends—basic channel protection stuff.

#gaysex's name alone ensured that it was a popular target for Robobot-style annoybots and clonebots. And Cevin could not cope well with abuse. He had an Achilles' heel that made him vulnerable to an insidious variation of one of the most common methods for assaulting a channel—the floodbot attack.

A floodbot joins a channel and proceeds to flood it—with garbage text, endlessly repeated insults, or random billowing storm clouds of data. A floodbot attack has a number of repercussions. First and most obvious, a floodbot makes it impossible for the regular members of the channel to communicate normally. Quick-fingered IRCers could set their accounts to ignore any text issuing from a floodbot, but such measures served only as stopgaps.

The IRC server protocol incorporated a built-in flood protection defense against floodbots. If any process that was connected to a channel generated more than a preset amount of output during a given amount of time, the server killed that process immediately.

One day, some #gaysex troublemakers discovered they could trick Cevin, the channel's own approved bot, into exceeding the server limit, simply by having multiple processes simultaneously ask Cevin for help. It didn't matter whether it was a group of humans acting in concert or a set of clonebots. The result was identical. Cevin would dump his three paragraphs of help information repeatedly, too fast for the server, and then get dumped himself.

"Three people message it 'help,' server kicks it off," says Pointer. "Completely worthless. Once everyone knew how to do that, there was no point in having a bot."

After Cevin had been knocked off the channel, of course, the channel attackers had free rein. Pointer saw a need and came to the rescue. He wrote a few lines of code for a generic helpbot that ensured it would never exceed the server limit, no matter how tempted. Cevin was left to concentrate on his channel protection duties.

Meanwhile, back in #gayteen, Pointer's other IRC haunt, a

raging power struggle had alienated most of the regular members of the channel.

"There were two sets of people who were always fighting for control," says Pointer. "And it kept getting nastier and nastier."

After conferring with some friends, he decided to stage a coup, kick out both warring parties, and install a combination channel protection/helpbot that could not be flooded out of the channel. He named this bot "Eggdrop." Eggdrop, the all-purpose, state-of-the-art channel protection bot, soon became the single most widely operated bot in IRC history, the defensive bot of choice for hundreds of beleaguered chat rooms.

Pointer, in the best tradition of Jarkko Oikarinen, made Eggdrop available to all. He asked for nothing in return, though he was always grateful for suggestions and bug reports. Pointer's main concern was that every channel have the right and ability to defend itself from those who would disrupt it.

It takes a bot to fight a bot, says Pointer. "We needed automatic protection against their automatic attack."

< >

Eggdrop was a smash success. Channel creators, desperate for peace, set up eggdrop bots, modified for local circumstances, in channel after channel. Valis, the motorcycle-bull-dyke bot in #gayteen, is an Eggdrop bot. So are Habanero and Jalapeno, two recent patrollers of #gaysex. Eggdrop's source code is available at multiple archival sites scattered across the Net, as are instructions for how to set it up and employ it to best advantage. And Pointer still tinkers with it, still adds new features and new defensive capabilities, when he finds spare time from his current job, working for a temporary employment agency that contracts out systems administration specialists to Silicon Valley companies.

But despite all of Pointer's success, he now has trouble logging on to IRC under his real name. He's persona non grata to nearly every IRC server administrator.

Shy and so soft-spoken that the clatter of cutlery on plates at the restaurant often threatens to drown out his voice, Pointer seems like an unlikely candidate to have his account banned all over IRC. But Eggdrop is the reason why. By 1996 the bot climate had gotten so bad on IRC that most IRC operators were banning all bots and all botrunners from server access on principle. They did not care that Pointer intended Eggdrop to be used solely as a defensive measure. Eggdrop was a bot. No bots allowed.

Pointer had sown the seeds of his own demise. By setting up Eggdrop as the crème de la crème of channel protection bots, he provided a target for every botslinging desperado in IRC to test their bot-fighting skills against. Where there is an Eggdrop bot, there is bound to be trouble.

There's an unfortunate irony to Pointer's story. As an altruistic hacker who devotes his code-writing skills to the public welfare, Pointer follows in the grand tradition of the Net's "gift economy." For years, Usenet, the Internet, and the Web have progressed according to a formula whereby individuals volunteer the fruit of their mental labor for the good of the entire Net, without expectation of financial compensation. In the gift economy, you are rewarded by the respect of your peers and the satisfaction of seeing your software adopted as a solution by the masses. For many netizens, it is a point of pride to live by the gift economy code. The principles of the gift economy are what keep chaos from overwhelming the anarchy generated by a decentralized out-of-control environment. On the Net, anyone who wants to contribute something can, and anything that solves a problem is welcomed with open arms.

But the gift economy works both ways. Free software is

available not just for solving problems. There's also plenty around for creating problems. As Pointer concentrated on building tools to make life in IRC better for all, so did some of his contemporaries focus on building weapons. And they too believed in the gift economy. They too made their warez available to all, whether as state-of-the-art clonebots, annoybots, floodbots, and collidebots—or as packages of "toolz" with revealing names like Gargoyle or Lice.

Like a horde of evil Prometheans, Pointer's dark-side counterparts raced each other to place ever stronger, ever more powerful, ever easier-to-use programs into the hands of the masses. This new generation specialized in making it easier for those who came after them to wallow in chaos.

And so, the era of the botwars began.

FLATTIE AND INT3NSITY

Flattie and Int3nSiTy hit #riskybus on the afternoon of October 23, 1995, seizing control of the popular Internet Relay Chat game channel after nick-colliding all the humans right off the Net. Like always, Flattie's guardbot had been watching his back all day, keeping an eagle eye out for any attempt by enemy bots to grab his channel ops.

A clonebot launched from a lagged IRC server broke through #riskybus defenses. Earlier that afternoon Flattie and Int3nSiTy had placed the clonebot on irc-2.mit.edu. They made their move when the server net-split, stranding one human #riskybus gameplayer.

Flattie had to kill the human—he was in the way. Meanwhile, the clonebot did what it was designed to do, spawning a mob of baby bots in quick succession with the nicknames of all the gameplayers currently on #riskybus. The IRC protocol forbids two beings with the same nickname, human or robot, from coexisting on a given channel at a given time. So when the server rejoined the

Net, all hell broke loose. The nicknames collided. Flattie and
Int3nSiTy ruled.

Kenrick Mock, the man behind the #riskybus game channel,
sighs when I ask him how often IRC botrunners attack his
game channels.

"Two, three times a day," he says. "Sometimes as many as
ten. It's a fact of life. It's like a big war zone."

It is the fall of 1995, and Mock, just months away from
taking a job at the Intel Corporation, is a computer science
graduate student at the University of California at Davis. A reserved young man, Mock evidences no apparent outrage at the
ceaseless bot hammering that his channels endure. He accepts
the relentless assaults as an inescapable IRC reality. Mock's
game channels are popular—popular channels get attacked.

Mock's channels, most notably #riskybus, deploy his own
specially designed bots that combine both game moderation
and channel protection functions. And top-notch protection
bots inevitably attract bot challengers—as Eggdrop's Robey
Pointer had learned years earlier. To defend one's channel is to
invite attack.

And so it is with #riskybus. On any given day, #riskybus
might be occupied by some twenty to thirty IRC gameplayers,
gathered in the channel to match wits with RobBot, #riskybus's
resident gamebot. RobBot moderates an imitation *Jeopardy*
game, asking questions, keeping score, deciding winners, and,
on the side, defending the channel. And even though RobBot
is no bot slouch and is constantly being improved, he isn't perfect. At least once a day, says Mock, the botrunners stage a successful channel takeover.

Their most common technique involves a two-part strategy.
First, they wait for, or force, a net split. On IRC a net split occurs when an IRC server loses its connection with the rest of

the IRC network, thus splitting IRC into fragments. Each of the new postsplit fragments believes it is the true IRC network, and has its own duplicate list of currently existing channels.

When the nets rejoin, the process of coming back up to speed takes time. By late 1995, fifteen thousand people are merrily chatting away in thousands of different channels. Every single IRC server computer needs to keep abreast of hundreds of megabytes of information detailing exactly who is in what channel, the ops status of each user, and so on.

IRC botrunners specialize in taking advantage of that window of opportunity in which a "lagged" IRC server struggles to catch up with the rest of the network. During such a window, they infiltrate a channel, receive ops, and seed the channel with collidebots. The collidebot's purpose is to generate clonebots bearing the nicknames of all the occupants of the channel they wish to take over. When the lagged server finally does catch up, the clonebots and the real occupants of the channel are momentarily in the same IRC virtual space, an occurrence that contravenes the laws of IRC physics. A nick collision results—the simultaneous expulsion of both clonebot and legitimate channel resident. The coast is then clear for the takeover artist to seize complete operational control.

According to Mock, the main IRC network, EFNet, splits scores of times, every single day. Opportunities for attack are legion. And indeed, the very next day after I speak with Mock, Flattie and Int3nSiTy, two members of the IRC gang REBEL-LiON, successfully raid #riskybus. Their triumph is utter, forcing Mock to cut a deal. In return for ceding control of #riskybus back to him, he gives them my email address.

So begins my descent into the subculture of the IRC gangsta—the scourge of IRC civilization. For months, a flurry of self-aggrandizing email messages from IRC botrunners with nicknames like MaytricKz and Synergist, Om and Seshon, and, most of all, Flattie and Int3nSiTy stuff my email box with tales of botrunning glory. Their boasts and brags paint a garish picture of a world where bands of teenagers congregate together in an ever changing kaleidoscope of savage alliances. Killa Fresh Crew, Toolshed, IRCuzi, DeSynk, the IRC Terrorists, Madcrew, Outbreak, the IRC Mafia: these gangs have their own secret channels, travel across the IRC landscape flanked by their personal bot posses, and constantly seek out opportunities to lay waste to IRC channels.

"REBELLiON," Int3nSiTy tells me, "is the best IRC takeover group out there. It consists of the best IRC hackers, and what we do is just pretty much piss people off and get revenge on channels that either we hate or that have channel operators in them that we hate. . . . We net-split, hack ops; we flood people off IRC, nick-collide people, and do what we have to to take over the channel."

Life moves fast in the IRC gangsta underground. Only a day after first making contact, Int3nSiTy asks me to excise all mention of the name "REBELLiON." As another former RE-BELLiON member later informs me, after a merger with the gang IRC Terrorists, a REBELLiON faction decides the group is too large and needs a purge. Now Flattie and Int3nSiTy belong to a new group—DeSynk. And a few months later, I learn, Flattie and Int3nSiTy themselves become victims of a purge.

"It's a fierce life," says Jonathan Hart, aka Synergist, another REBELLiON/DeSynk gangbanger. "There's a lot of espionage between groups—there are spies, backstabbers, extortion, scapegoating, lying, stealing, and a lot of colliding."

Hart offers me an eloquent synopsis of the internal psychology of the IRC gangsta: "We operate much like a city gang would. We have a hangout, we have enemies, we have other gangs against us. We have friends, connections, and countless resources to finish the job. We aren't computer nerds with thick glasses, either. I wear contacts.

"A lot of hackers have different motives," says Hart. "Some just want to cause mischief, others want to feel 'cool,' while many see it as a high-tech game of war and conquering, over-throwing the operators of a channel, and claiming it does give you an adrenaline rush. I remember my first real takeover. It was in #hottub. It had about forty users in it, and they were all ops. I collided every one of them and split a guardbot in to take the channel. I was local-killed, and they soon got it back, but in those few moments I felt the biggest power trip of my life. A lot of hackers are addicted to that power trip and don't let go."

Some IRC hackers do use their power for good. Flattie spends part of his online time as a protector for #biteengirls, a channel for bisexual girls that, Flattie says, "is taken over so much it's not funny." But that's not his main gig.

"Raiding channels is as fun as anything on IRC," Flattie tells me. "I mean, who wants to just chat all day long and just become a couch potato? For me, if I'm banned from a channel, I tend to want to take that channel. Being banned is like red to a bull."

"My bots are evil," adds Int3nSiTy. "I like to get people mad. My account is banned from nearly every server on IRC. It's almost time for a new one."

< >

Initially, IRCbots were useful tools or curiosities or play-things—superlative examples of the cyborg synthesis occur-ring on the human-computer interface frontier. Inherently neither good nor bad, IRCbots reflected the impulses of their owners, and many of them lived long lives as upstanding IRC citizens. But by late 1995, much of the IRC community regarded all bots as menaces to society. A mighty bot backlash began, the inevitable result of the botwar era. Even some early infamous botwriters, legendary for their contributions to the bad bot archives, renounced botrunning after becoming appalled at the chaos that raged through IRC in the years 1993–1995.

"My stance on bots has changed dramatically," says "Hendrix," the author of a particular class of annoybots that make the ancient Robobot look like a ninety-pound stripling bot. "I'm now pretty anti-bot myself. I mean, I still get ops from bots and occasionally run one of my own, but I think the proliferation of bots is just way out of hand. It seems like within a week of IRCing, every new user wants to get three bots of his own to run to protect him."

The backlash led to a concentrated crackdown. Acting individually, a majority of IRC server administrators—ultimately, the only authority of any kind in IRC—banned all bots from their servers. Some cited congestion as their reason. Every bot connected to a server usurps the place of a potential human. An IRC server can accept only a limited number of connections, and the current state of chronic IRC overcrowding means that most servers are permanently maxed out. As far as server administrators are concerned, bots have no inalienable rights. Humans always come first.

But congestion wasn't the only reason. Bot-created anarchy drives the server administrators crazy. What they cannot control must be eliminated. And by late 1996 they had succeeded. The bot crackdown cleaned up IRC, at least as far as anti-bot vigilantes were concerned. In 1996, IRC had far fewer channel takeovers and much less overall disruption. Technical changes to the main IRC server protocol contributed to some of the positive change, as did a sprinkling of new IRC subnetworks established by IRC veterans fleeing the chaos of the old IRC world. But the most important element of the crackdown was the vigorous anti-bot policies put into place at individual servers.

These anti-bot regulations raise questions that cut to the core of the human-bot relationship. Defining what is "human" and what is "bot" in a software environment like IRC is a task fraught with contradictions. There is no clear dividing line. Bots and humans go together. Does banning bots mean banning all the not-quite-autonomous tools that give humans bot-like powers? Do all scripts, for example, qualify as warscripts? The adrenaline-crazed IRC gangstas did not confine themselves to the companionship of autonomous bots. They scurried around IRC with arsenals of code auxiliaries. Again, how does an administrator decide what is an allowable tool and what is a forbidden weapon?

The answer, of course, is that any tool can be used as a weapon—the line is drawn by interpretation and motive, not by any intrinsic quality. Tool user ethics—programmer ethics—make the difference.

But on IRC they didn't make any difference at all. The ethics of the many proved no match for the unbridled egos of the few. The upshot? Good bots lost out in the battle. Today's IRC has far fewer servicebots than it once did. Gamebot runners like Kenrick Mock require close relationships with server administrators just to be allowed to run their bots at all. Robey Pointer can hardly show his face.

There's a moral to the story. It is easy to dismiss IRC as an elementary school playground—a park full of high-pitched shrieks, water fights, and spirited name calling. After all, the average participant is quite young, and as Jonathan Hart notes, "What can you expect from a bunch of twelve-year-olds?" Just wait until all they all grow up, and the problem will take care of itself. They'll learn how to be ethical human beings.

And, surely, their juvenile antics have no bearing on what is going on in the rest of the Net, where serious adults are engaged in serious business. But is IRC really an elementary school at recess or is it more akin to a staging ground for a cyberspace version of *Lord of the Flies*? Isn't the example of IRC a warning, a fable outlining what happens when you let loose a bunch of kids without supervision from a principal or recess monitors, without the deterring threats of detention or ruler-delivered raps on knuckles?

IRC couldn't handle its unchained wild things. In an anarchic, decentralized, no-rules setting, the admonitions of bot etiquette had no teeth. And so bot excesses led to the squelching of all bots. The glory of the Cambrian explosion is now a faded memory. The myriad IRCbot species are on the verge of mass extinction, with only a few life-forms struggling for oxygen in out-of-the-way tidal pools.

Could IRC be the canary in the coal mine for the Net ecology, for the once mighty gift economy? Certainly, Robobot and its fellow bad bot brethren are not welcome role models for intelligent agent behavior. But are they the inevitable result of unrestricted bot evolution?

Other Net-based societies—in Usenet, in isolated MUDs, and on the Web—are now being forced to face up to the kinds of problems caused when humans run amok with software that allows them to exert power online. Such online communities are increasingly vulnerable to the ill feeling spawned by schisms of race, religion, and ideology. The arrival of ever more powerful bots threatens to create an accelerating crisis, to widen those schisms irresponsibly, anonymously, and automatically.

The absence of government has an inevitable corollary: anarchy breeds unbridled discontent. IRC—small in size, juvenile in behavior, relatively unsophisticated technically—offers a powerful counterargument to those who trumpet the glories of intelligent agents in the information age. As bots and agents are used more often for serious work, they too will be abused. Already, bots and agents are being employed on the Web by individuals who couldn't care less about the long-established customs and habits of the Net's original inhabitants. Only now the stakes are much bigger. Hundreds, if not thousands, of entrepreneurs and hackers are hard at work devising programs that make the potential power of an individual in cyberspace immense. And they are doing so for profit, not for fun. Will they agree to voluntary limits? Or is bot mayhem, as Kenrick Mock acknowledged, an inevitable fact of life? Will the Net be forced to crack down on rogue intelligent agents and out-of-control Web robots, just as IRC cracked down on IRCbots? Will some middle ground be found that will allow bots and agents to flourish peacefully?

Or are we poised for an era of botwars that makes what happened on IRC look, indeed, like child's play?

6

RAISING THE STAKES

DATABASE RAIDERS

Alex Cohen doesn't like what he sees. The intruder is snooping through the McKinley Corporation's computers, leaving footprints a mile wide that record every access of the company's Web servers, the computers that house all of McKinley's Web-based documents. The invader repeatedly hammers on locked files containing password-protected data that is off-limits to the general public. And it does so in rapid-fire succession, spawning new requests for documents nearly every second.

Cohen, chief technical officer at the McKinley Corporation, immediately fingers the bull-headed visitor as a bot. No human could or would manually request so many Web documents in such a short period of time. And no human would be dumb enough to continue attempting to retrieve files that are so clearly inaccessible. Such high-speed stubbornness is the calling card of a Web robot, a fact Cohen, a robot author himself, knows well.

The robot incursion riles Cohen. The robot has blasted

right by Cohen's digital Do Not Enter sign, a special file containing instructions for all visiting robots. If the robot had read those instructions, it would have known better than to attempt to retrieve the private files, and it would have been ordered to restrain its quick-trigger document requests to a more manageable rate. A Web robot accessing hundreds of documents in a matter of minutes can bring a Web server to its knees, so overloading its capacity that it becomes useless for humans and robots alike. Cohen considers any robot that ignores his Do Not Enter file, which follows all the prescriptions of the Robot Exclusion Protocol, to be in egregious contravention of established Web robot manners.

Still, in the spring of 1995, new Web robots are emerging from the Web woodwork with increasing regularity, and not all robot authors are familiar with the Robot Exclusion Protocol, a voluntary standard that has no official backing. The Web is young. Mistakes will be made. Ordinarily, Cohen is prepared to be forgiving.

But not in this case. This Web robot has been launched by someone who should know better, from a computer owned by a direct competitor of the McKinley Corporation. The target of its requests—that password-protected database—constitutes the McKinley Corporation's stock-in-trade: its painstakingly compiled catalog of hundreds of thousands of the most useful, informative, and entertaining homepages on the World Wide Web.

The McKinley Corporation is in the Internet directory business, or at least it will be soon, once it finally unveils its guide to the Internet, Magellan. McKinley plans to strike it rich by capitalizing on the single most glaring shortcoming of the World Wide Web: its utter lack of organization. McKinley's founder, Christine Maxwell, has already attempted to tap that vein by coauthoring one of the first hard-copy Net directories,

New Riders' Official Internet Yellow Pages. But Maxwell, a former professional "information broker" and the daughter of the late British publishing tycoon Robert Maxwell, has abandoned her father's hidebound world of print. Her bet is on the Web.

She isn't alone. Scores of other entrepreneurs have recognized the same fact—people want a map of the information superhighway. Yahoo!, one of the first Internet directories, founded in the spring of 1994 by a couple of Stanford graduate students, is leading the way. By March 1995, the Yahoo! Web server is registering ten million hits a week—ten million computerized requests for information. Other catalog and search engine sites see similar results, if not quite in the same range.

Hundreds of volunteer Web surfers and employees compile the McKinley Corporation's database of rated Web pages (Web sites reviewed for clarity, usefulness, and organization). Alex Cohen's Web robot, the Wobot, compiles an even larger database of unrated pages. As 1995 gathers steam, all across the Web, similar searchable databases of information concerning the Web itself are rapidly becoming cyberspace's most precious commodity. And no one is better aware of the economic importance of such information, of course, than another Internet directory company.

Cohen isn't overly worried about protecting the information. His database is structured in such a way that a normal none-too-bright Web robot will not be able to make any sense of how the information is organized, even if it manages to circumvent the password protection. And if he wants to, Cohen can block all requests to his machines originating from the offending company. He even contemplates setting up a trap to ensnare unwary Web robots, but refrains. He isn't absolutely sure whether the trespassing bot has been targeted directly at McKinley in a brazen attempt to steal its entire URL catalog or whether the bot just happened to wander by McKinley's home-page while on a general Web walkabout.

But Cohen still has reason to be annoyed. Whether this particular robot has indeed been engaged in industrial espi-onage is almost beside the point. Any company using Web ro-bots for profit should be aware of guidelines such as the Robot Exclusion Protocol or the more general rules of etiquette re-garding Web robots. If the principals of such a company don't care about such considerations, then so-called Web standards like the Robot Exclusion Protocol have no teeth, even as the ever-rising status of bots as vital information industry cogs un-derscores the necessity for some kind of mutually agreed upon rules of behavior.

Alex Cohen is the kind of man who talks at the speed of light and still can't tell you everything he wants to say. In con-versation, he thrusts and parries over an intellectual landscape that includes French poststructural theorists as often as it does John von Neumann's explication of cellular automata. For Co-hen, an ace programmer who moonlights as an academic spe-cialist lecturing on the treatment of technology in films such as *The Terminator, A Clockwork Orange,* and *Blade Runner,* the po-tential ramifications of unethical bot use summon up visions of a dark cyberpunk future. A future of bots battling bots in a no-holds-barred free market. A Gibsonian nightmare popu-lated by real live Maas-Neotek corporations employing bot surrogates to ferret out private information about citizens and promote corporate propaganda campaigns.

Push has come to shove in cyberspace. The economic mat-uration of the Web has generated real conflict—clashes be-tween inhuman proxies in a digital domain. Forget the IRC botwars, the Barney plagues, and bum bot bungling. Such kindergarten antics pale against a backdrop of high-stakes gambling, where bots and humans are players shooting for big bucks and palpable power. Where that game will end—with the banishing of all bots, as has happened in IRC, or in the chaotic conflagration of a bot-induced info-Armageddon, or in some stable Elysium of bot and human harmony—is an unanswerable question, for now. All one knows for sure, from the evidence in Alex Cohen's Web server that May morning in 1995, is that the game has begun.

SPIDERLAND

In the Web's vernacular, Web robots are commonly referred to as spiders or Web crawlers. The insect metaphor is apt in some ways, misfires in others. Web robots are true creatures of their environment. Their purpose is to explore the Web itself, not to engage in silly conversations with humans or to protect online chat room turf from harassment and assault. A Web robot scurries from hyperlink to hyperlink, weaving a mighty index from a million shining strands of Web documents.

Except that, technically, a Web crawler doesn't crawl. There is no scurrying, scuttling, or silken-thread shimmying. A spider is a stationary computer program that sends requests for documents to Web servers. Once having retrieved a file or document via the Internet, it can examine it for keywords or for embedded links represented in hypertext markup language (HTML) as URLs—computer-comprehensible addresses of Web documents. If so instructed, the spider can then send out requests for the new URLs and retrieve a fresh set of documents. And so on, in an endlessly branching tree throughout the Web.

The Web welcomes all bots. Chatterbots like Eliza and Julia have taken up residence on Web pages. IRCbots, frozen in suspended animation, are accessible in Web-based archives of code. Gamebots lurk, waiting to do battle, in online gaming networks increasingly interwoven with the Web. The Web, irresistibly connecting and binding together all the communities of cyberspace—MUDs and bulletin boards and chat rooms—is the reconvergent Pangaea, the mother continent where all bots will one day roam.

But right now, Web robots are the dominant Web life-form. They are the strings of code most suited for navigating the Web's hyperlinked reality, the focus of sustained academic and corporate research, the first real precursors to the myriad of agents poised to descend on the Net. Though they are devoid of most familiar bot characteristics—speech, personality, individuality—their crippled personas mark only their immaturity, not their full potential. Web robots have just cracked the shell of their nurturing eggs—they are far from the fully evolved bots that they will become. Even now, Web robots like AutoNomy's canine searchbots are being endowed with rudimentary natural language capabilities and dabs and sparkles of character. As Web robots gradually incorporate the trappings of more personable bots, they will point the way to the emergence of a superior bot species in which all the characteristics of the Net's many bots combine in one form. Even in their current utilitarian guise, as they scavenge the Web tirelessly for dribs and drabs of information, they signal the shape of things to come.

< >

In May 1993, Matthew Gray, a physics student at MIT, wrote the first widely recognized Web robot—the World Wide Web Wanderer. Gray's goal was simple—to answer one question. How big is the Web?

Back in January, Gray had begun investigating "this Web thing that some guy in Switzerland had set up." Tim Berners-Lee, a researcher at the CERN High Energy Physics Lab in Geneva, had invented HTML, the programming language that allows users to write documents that can be linked together via the Internet. Berners-Lee soon became known as the father of the World Wide Web, that sprawling subset of the Internet constructed out of all those interlocking HTML documents. Gray's research in high-energy physics made him naturally interested in this new technique for accessing scientific documents, even though in January 1993 there were only a "couple of dozen Web servers" on the Net.

Today's Web surfers would be lost in the landscape traversed by Gray. Hyperlinked documents had no formatting, no pictures, no sound files, no cutesy animations—just straight, unadorned text. Plain numbers enclosed by square brackets represented links. Nearly all the information available was of interest only to physics researchers.

"It was very poor," says Gray. "But it worked."

It worked well enough for Gray to decide he wanted to set up his own Web server. And to dream of a "central repository" or listing of all other available Web servers. But no one then knew how many Web servers existed, and Gray wasn't thrilled at the prospect of manually clicking on every hyperlink he could find, even in those early days. Clearly, this was a job custom-made for a bot.

"I wrote a robot in Perl," says Gray, "and I set it loose."

In May 1993 the Wanderer counted 100 Web hosts—unique "domains" from which one can retrieve Web pages and which can harbor thousands of individual homepages. By November the number of hosts, or Web servers, reached 600. Two months later, in January 1994, there were 2,000. Two years later, 100,000. By June 1996, 200,000, and as of January 1997, the number had doubled to more than 400,000.

The Web's growth curve—in 1994, the Web doubled in size every two months—puts all other Net-related growth curves to shame. It signals the transformation of the Internet from a playground for academics, geeks, and subsidized college students into a mass-media phenomenon. The Web made the concept of cyberspace, long a staple of recklessly speculative science fiction, a part of daily life, accessible from coffee shops and airports, laundromats and bars. By 1996, Web page addresses appeared on billboards and in television commercials, popped up in comic strip panels, and adorned millions of business cards. Executives at corporations both large and small stood up and took notice, asking themselves how they could integrate Gray's "Web thing" into their operating strategies.

Growth came so suddenly and so exponentially that it overwhelmed any feeble human attempts to make sense of it. A mighty paradox ensued. What practical benefit could be derived from an info-bonanza when no one knew how to find the information they needed? In May 1993 a person could connect to the Web and in a matter of hours tour every new Web site created since he or she last logged on. But by the fall of 1993 no one had enough time to visit all new sites, even if they were so inclined. Real-world time demands got in the way.

At the University of Washington, Brian Pinkerton, a graduate student in the computer science department, had a thesis to write—a major headache involving the concoction of a set of gene-sequencing algorithms useful in molecular biology. That fall, says Pinkerton, he found it "incredibly frustrating" not to be able to wander the Web at his fancy, like some of his less busy colleagues.

"They had all this time to surf around the Net and find this interesting, cool stuff," says Pinkerton, who looks more like a stereotypical California surfer than the standard monitor-addled technogeek.

So Pinkerton wrote his own Web spider—WebCrawler.

WebCrawler represented a major advance over the World Wide Web Wanderer. WebCrawler retrieved entire Web documents and set them aside in a database. Anytime he wanted, Pinkerton could take a break from his thesis and punch a keyword into his home-cooked search engine. The search engine program ran through his index and generated a list of possibly relevant URLs. Pinkerton connected his search engine program to his Web browser, enabling him to go straight to the Web from his keyword searches.

Now Pinkerton had a new thesis topic—WebCrawler. He had climbed the biggest wave ever to roll through the Net, and a roiling crowd soon joined him. In computer science departments and research institutions all over the world, Pinkerton's colleagues reacted to the uncertainties of the Web's information explosion by unleashing a swarm of creepy-crawly critters into cyberspace.

Carnegie Mellon's Michael Mauldin abandoned TinyMUDding and Julia and set to work on Lycos (derived from the Greek word for "wolf spider"), an indexing spider similar to WebCrawler. Lycos exhibited little in the way of personality— no quips about ice hockey or guinea pigs issued from its code engine—but that didn't matter. Chatterbot-style gabbing could not compare to the challenge offered by the Web.

While Pinkerton honed WebCrawler and Mauldin perfected Lycos, Jerry Yang and David Filo of Yahoo! employed robots to help them construct their Web catalog, programming spiders to regularly peruse popular what's-new lists maintained by some of the biggest Web publishing sites. It began to seem as if every Web-related info-problem called for the same answer: build a bot.

In the classic tradition of computer science, when faced with information management problems beyond human control, hackers conjured up a horde of daemons to come to the rescue. As 1994 progressed, Web robots proliferated like mad, swarming into cyberspace with the rapacity of a rampaging mob. They knocked on every Web server door, struggled to penetrate every nook and cranny of the Net. And as the Web exploded in size, so did Web robots burgeon in strength, speed, and cunning.

In the Web's wake, the insular worlds of MUDs and IRC rooms now seemed like a mere prelude to the big show. These new bots were neither toys nor diversions. They were vital tools. But the same mischievous spirit that pervaded every niche of the computer world still ran strong. The new additions to the bot family tree thrived in goofy taxonomic splendor: MomSpider, tarspider, and Arachnophilia; Scooter and Aretha, Python and Peregrinator, JumpStation and churl; webfoot and Wobot, HTMLgobble and Websnarf.

Still, as zany as these robots pretended to be, their arrival trumpeted a major evolutionary leap forward for all bots. These bots were the true heirs to Maxwell's Demon—autonomous programs performing useful services for their human masters, creatures of information navigating seas of data. Instead of combating the entropy of the universe, they battled the entropy of info-overload.

Make order out of chaos—so goes the Web robot mandate. Without them, the dark continent of the Web will remain an impenetrable jungle forever. With them, civilization becomes possible.

ROGUE SPIDERS

Programmer Thomas Boutell wears many hats. He maintains the frequently-asked-questions file for the World Wide Web, created a popular technique for compressing graphic images into small files, and authored a book on programming Web servers. He also administers the World Birthday Web, a Web site where Web surfers can register their birthday in the hopes of getting virtual birthday cards in their electronic mailboxes, come that happy day. The World Birthday Web has an individual Web page for every single person registered there. There are thousands of such pages, and those pages are in turn linked to thousands more off-site pages.

One day, a Web robot launched from a Norwegian Web server (or multiple servers—Boutell still isn't sure) began clicking, repeatedly and mindlessly, on every single one of the World Birthday Web's pages. The sudden rise in traffic dragged all the Web servers at Boutell's Internet service provider (the company hosting his Web site and connecting him to the Internet) to a screeching halt—the multiple demands for Web pages in such a short time clogged up all available bandwidth. Boutell's provider had no recourse but to "lock out" an entire Norwegian Internet "subdomain"—thus forbidding any further requests from any computers within a whole region of Internet addresses.

Whoever had launched the Scandinavian Web robot probably wasn't acting maliciously, manifesting some deep hatred for birthday celebrators. Web robots can be used in purposeful attacks (referred to, in the trade, as denials of service), but motivated assaults in the early days of the Web were rare. Much more likely, the blundering Web robot was an accident, a programming error by a Web novice unaware of the horror that can be precipitated by a badly designed arachnid automaton.

Accidents, however, especially in the earliest days of the Web, were all too common. From the point of view of a machine being accessed by a Web robot, a single hit generated by a robot is indistinguishable from a hit generated by a real live human being. But a badly behaved Web robot can do far more harm than the most frenetic of wildly clicking humans.

An example of an especially stupid robot trick occurs when a bot attempts to retrieve Web pages that have mutual links. If the right precautions haven't been taken, the robot can retrieve the same two pages endlessly, chewing up bandwidth and processing power. "Black holes" created by pages that automatically generate new documents at the click of a button present another robot danger. The last page hit by the robot creates another page, which is then hit in turn—a nice and tidy definition of infinity. Robots also have a tendency to attempt to retrieve files that they cannot handle—video or sound files, or huge compressed documents that can easily reach book length and might take hours to retrieve over a narrow bandwidth connection.

As hacker after hacker rushed unfinished and imperfect Web robots out into the Web, webmasters—the administrators of Web servers—became steadily more agitated. Rogue spiders kept busting into their carefully tended Web gardens, trampling over the fruits and vegetables and messing up irrigation systems, all without so much as a by-your-leave. But there was no one to complain to, no Robot Infractions Bureau to dole out tickets to reckless robots. Webmasters had to come up with their own line of defense.

In 1993 the first webmaster to spring into anti-robot action was Martijn Koster, an engineer at a British software company named Nexor. In person, Koster is a tall, friendly man who enjoys a wry chuckle. His online persona, however, can come across as didactic and stern, to the point that his ferocious robot denunciations at one time earned him the epithet of "rabid anti-robot attack dog."

As Nexor's webmaster, Koster gained firsthand experience of Web robot abuse early on. "I just happened to be the one of the first people to get hit by a bad robot and decided to sit up and do something about it," says Koster.

Koster devoted himself to becoming a one-stop clearing-house for information on Web robots. He set up a mailing list devoted to robots, as well as a Web page with a regularly up-dated list of all active robots. He personally contacted robot authors such as Brian Pinkerton and Michael Mauldin to in-form them of the potential consequences of their actions. Most important, he conceived of and authored the Robot Exclusion Protocol—the first stab at an ethical mandate for robots on the Web.

The key to the Robot Exclusion Protocol is the robots.txt file—Alex Cohen's digital Do Not Enter sign. The robots.txt file contains all the information a visiting robot needs to know to act properly. But the protocol also sets more general guide-lines: robots should limit themselves to a particular speed, ro-bots should avoid certain kinds of files, robots should search through trees of Web pages in a specific order so as not to place undue burdens on servers, and so on.

The Robot Exclusion Protocol makes a great deal of sense. But it has one serious drawback: it is utterly unenforceable. There is no immediate way to prevent a noncompliant robot from having its way with a Web site. If the robot has already been identified as a troublemaker, the domain that it originates from can be blocked. And if a Web server has been equipped with top-notch robot detection defenses (algorithms that watch for moments when requests from a single domain start coming in too rapidly to be human in origin), it may be able to stop an assault already in progress. But there's still no way to prevent a hit-and-run attack. There isn't even a way to ensure that an unidentified robot stops by the robots.txt file to read the house rules. For the Robot Exclusion Protocol to work, ro-bot authors have to design their robots to visit the robots.txt file voluntarily. Robot authors have to support the protocol.

And at first, those authors who learned about the protocol

generally did support it. In 1993 and 1994, Koster's invocation of robot ethics fell on receptive ears. Most Web robot authors professed themselves concerned about "programmer ethics"; the community of Web-savvy programmers was small enough that peer pressure could have an impact.

Not that it really needed to. Most of the major robot authors knew each other, meeting regularly at international conferences devoted to the Web. Pinkerton, Koster, Mauldin, Gray, and a few others even formed an informal robot authors group, Braustubl, named after a pub in Darmstadt, Germany, where they met once for beer.

As scientists with the primary goal of dispersing information, the Braustubl band and its contemporaries took Web resource conservation issues seriously. They did not want to impose undue loads on the Web's infrastructure or trespass where unwanted. They were card-carrying members of the gift economy—the kind of upstanding cybercitizens who had made the Net a smashing success by selflessly donating their time, energy, and skill. Like IRC's Robey Pointer, they intended their robots to contribute to the general welfare: they placed their Web indexes and catalogs on Web servers open to the general public, and they invited any and all to enjoy the fruits of their Web robotic labors.

Their cooperation ensured the Robot Exclusion Protocol's immediate triumph. By the end of 1994, reports of robot abuses to Koster's Web robot mailing list plummeted. Not only did robots regularly check for robots.txt files, but they also adhered to common standards of robot decency as they did so. Scrupulous webmasters examined their log files regularly— and on those increasingly rare occasions when a misconfigured robot went awry, they made immediate attempts to educate the transgressor, usually with success.

Koster's crusade to inculcate an ethical standard for robots seemed to be working, providing a beautiful example of a community governing itself through consensus and voluntary cooperation. Throughout 1994 and on into early 1995, the Web robot gift economy machine ran like a well-oiled sportster, hardly even needing a pit stop. Then the wheels fell off.

THE NASTYGRAM

This www server has been under all-too-frequent attack from "intelligent agents" (a.k.a. "robots") that mindlessly download every link encountered, ultimately trying to access the entire database through the listings links. . . .

We are not willing to play sitting duck to nonsensical methods of "indexing" information. This server is configured to monitor activity and deny access to sites. . . . Continued rapid-fire requests from any site after access has been denied . . . will be interpreted as a network attack; and we will respond accordingly—without hesitation, and without further warning.

—Robots Beware, Web page at
xxx.lanl.gov/RobotsBeware.html, July, 1996

One summer day in 1996, a computer programmer named Aaron Nabil made a fateful decision. He took some URL collection algorithms he had been hacking together for a "next-generation search engine" out on the Web for a test drive. In his own mind, the excursion was an innocent exercise. He merely wanted to make sure some modifications to his code would work as intended.

Unfortunately for Nabil, he chose the wrong racetrack to rev his engines on. He targeted his algorithms at a list of URLs from the e-Print archive Web site. To Nabil, the archive seemed an eminently sensible destination. One of the biggest and oldest Web sites in the world, the e-Print archive is a vast repository containing thousands of scientific documents, laced together in an intricate network of complex HTML code. If Nabil's algorithms could unlock the mysteries of the e-Print archive, then they could confront the entire Web with confidence. No site would be safe from their prying fingers.

But mere seconds after Nabil's program began slithering into the e-Print archive, its presence set virtual alarm bells clanging all over *xxx.lanl.gov,* the Web server that hosted the archive. In marked contrast to run-of-the-mill Web servers, *xxx.lanl.gov* was armed to the teeth with a battery of ultrasensitive trip wires and defensive emplacements, all configured to respond to exactly the kind of automated inquiry initiated by Aaron Nabil's program. The e-Print archive regarded most Web robots as barbarians at the gate. Cauldrons of boiling oil and a piranha-packed moat awaited any assault.

The "Robots Beware" page, one short hop from the e-Print archive main page at *xxx.lanl.gov,* made no secret of its antipathy to mechanized visitors. At the top of the page, a skull and crossbones glowered at all visitors, flanked on the right by a fat red X plastered over a small picture of Commander Data, the Star Fleet Academy android officer in TV's *Star Trek: The Next Generation.* The screen text read like an unbending ultimatum: robots were not welcome at *xxx.lanl.gov,* period. The act of unleashing a robot that repeatedly disregarded the Robot Exclusion Protocol was tantamount to declaring war on the Web server. And the e-Print archive, secure in its home base at the Los Alamos National Laboratory, would not shy from responding in force.

Before he launched his program, Nabil, the owner of a small research and development firm, had no idea that the "Robots Beware" page existed. But his enlightenment was fast in coming. Shortly after initiating his algorithmic venture, he received an email from one Paul Ginsparg, the administrator of the e-Print archive. In the letter, which Nabil later labeled a "nastygram," Ginsparg harangued Nabil for "attacking" the *xxx.lanl.gov* Web site with a robot.

"He accused my 'robot' of violating the 'robot guidelines,' " wrote Nabil. "He asked that I 'cease and desist.' "

Nabil, who had indeed failed to "respect" Ginsparg's robots.txt file, acceded but then personally paid a call to the *xxx.lanl.gov* Web site. There, to his shock and dismay, he discovered the "Robots Beware" page. And if the skull and crossbones wasn't alarming enough, he was even more astonished at another page he found linked at the site, emblazoned with the title "seek-and-destroy." The seek-and-destroy page purported to demonstrate the automated return attack that Ginsparg planned to unleash against foreign sites operating robots that disobeyed the Robot Exclusion Protocol. "Click here," said the page, and be annihilated.

The nastygram was just the first step in Ginsparg's defense procedure. Had Nabil ignored that warning, each subsequent automated request for a Web page generated by his Web robot would have resulted in a retaliatory barrage of rat-a-tat email messages—"a storm of mailbombs" designed to inflict pain and suffering on offending Web servers.

< >

Paul Ginsparg is a man fond of inflammatory statements like "the problem with the global village is all the global village idiots." His take-no-prisoners approach was no anomaly. If ever an award is given for the honor of being Robot Enemy Number One, Paul Ginsparg would be an odds-on favorite to ascend the podium.

Ginsparg hated robots for two main reasons. First, he had simply been around longer than almost anyone else on the Web (he founded the e-Print archive in 1991, well before the Web formally existed), and as the maintainer of one of the largest collections of documents on the Net, he had had to contend with more badly designed robots than most Web site administrators. Second, the nature of his site made robot incursions especially troublesome. Most of the documents in the e-Print archive are very large files stored in digitally compressed form. Indexing robots cannot normally decipher such files, and the act of retrieving a considerable number of them in quick succession is guaranteed to crash the e-Print archive Web server.

Ginsparg's experiences made him a partisan of the Robot Exclusion Protocol. His own log files proved that after the first rash of rogue spider abuses broke out in 1993 and 1994, widespread promulgation of the protocol had stemmed the tide of robotic malfeasance. Ginsparg himself, always quick to track down each new robot to its source, had played a significant role in ensuring that robot authors knew about protocol prohibitions.

But the Robot Exclusion Protocol was no panacea, and Nabil's misadventures were just one trickle in a new wave of robotic problems that challenged the efficacy of the standard. By the summer of 1996, stupid robot tricks were back on the upswing, and Ginsparg, a crotchety man to begin with, had had enough. No more passive waiting around for yet another robot to come by and mess up his operations. He intended to fight back. He readied his storm of mailbombs.

Even though Aaron Nabil obliged Ginsparg's "cease and desist" order and thus avoided cyberwar escalation, Nabil felt wrongly picked on as a target for Ginsparg's mutually assured destruction deterrent. Nabil did not even consider his set of URL collection algorithms a robot. In his view, they were merely an automated version of a human act, and that did not qualify them as an autonomous robotic entity. He rejected the authority of the Robot Exclusion Protocol. And he concluded a post to Koster's mailing list with this question: "Are the 'robot guidelines' no longer 'guidelines' but 'rules' and are these rules applicable to all forms of automated access, even if they aren't robots?"

Nabil's question had an effect on the mailing list subscribers akin to a spark flying into dry kindling after a year of drought. The list, normally a forum for mundane inquiries about indexing strategies and other technical arcana, exploded into the most virulent flamewar in its three-year history. Among the questions debated were straightforward ones: What defines a robot? What should be included in the robot guidelines? Who should be expected to follow them? If only 5 percent of all known Web sites had robots.txt files, who could claim that the Robot Exclusion Protocol should be treated as a standard?

But more incendiary issues also arose. One robot author revealed that hidden in the source code to the *xxx.lanl.gov* main page was a list of words—*sex, adult, nude, playboy, porn, porno, erotica, pornography, erotic, gay, penthouse, pussy willow, bondage, girls, xxx, nudity, nudes, hustler*—that only a robot could read. Was Ginsparg loading his site with commonly chosen search engine keywords in order to ramp up the number of hits? Robot authors and search engine operators loathe such a tactic, known as spamdexing.

Some discussants wondered what business a government-funded Web site had in threatening costly retaliatory measures against private citizens (Ginsparg worked at Los Alamos, which in turn is administered by the Department of Energy). The plot thickened when an unknown hacker, apparently pretending to be Paul Ginsparg, posted a long, positive review of the e-Print archive to the mailing list.

The nastygram spat was one for the bot annals. In a single

stroke, it brought into stark relief a host of complex issues sim-
mering beneath the uneasy surface of robot technology. When
did automatic exploration become automatic attack? When
did loose community standards become rigid rules? What
constituted appropriate robot behavior and what could be
condemned as unacceptable anti-robot escalation? As one
commentator observed, the nastygram affair was "a wonderful
mixture of government waste, automated server attacks, robot
paranoia, false keywords to generate hits to a Web site, poor
robot design, and the 'voluntary' nature of the robot exclusion
standard."

< >

As part of the nastygram aftermath, Paul Ginsparg altered
his Web site to remove the offending fake keywords. He also
toned down his rhetoric. But even as his concessions quelled
the *xxx.lanl.gov* furor, the nastygram debate exposed a fundamental division in the Web techno-cognoscente ranks. The
programming elite, who had hitherto seen themselves as part
of a community that, albeit fractious, still shared the same
overall goals, were now separating into distinct factions. Webmasters and robot authors saw the Web in profoundly different
ways. Webmasters operated from positions of power over their
own domains and eyed askance anything that infringed on
their ability to govern those domains as they best saw fit. Robot authors, in contrast, saw the Web as a huge public space
open to all, a vast wilderness crying out for exploration and
harvest. Freedom versus power, curiosity versus privacy—the
rifts brought into focus by the nastygram incident have been
widening ever since.

The gift economy consensus—that tacit agreement among
the technogeeks that all of the Net's problems could be

solved to the public benefit through altruistic and volunteer sacrifice—was breaking down. The constant struggles at *xxx.lanl.gov* were not the only fault line. Even before the nasty-gram episode, Bruce Krulwich, a researcher at Andersen Consulting, had written a Web robot/agent called BargainFinder. If given the name of a music CD, BargainFinder would visit a succession of online music stores and return a list revealing the CD's price at each store. But most of the music stores soon blocked BargainFinder. Some of the store owners complained that the BargainFinder bot accessed their sites so often that regular customers couldn't get in. Others simply wanted to avoid being compared with the competition. Again, private interests clashed with public desires.

In an emotional post to the Web robot mailing list, where robot authors and webmasters convened together on a daily basis, Rob Hartill, the administrator of the Internet Movie Database, a massive resource containing movie reviews and filmographies for thousands of films, offered a standard web-master's complaint: "Robot authors should follow guidelines and respect site owners' choices [as to] what should be hit automatically and what shouldn't." But instead of heeding the established standard, Hartill claimed, an increasing number of Web newcomers had arrived, armed with the attitude that "robot owners can do whatever they damn well feel like and it's the job of a site owner to build a fortress of safety systems to prevent the robots from causing damage."

Robot authors naturally took offense at Hartill's categorizations. First, they refused to recognize the authority of webmasters to decide what constituted proper robot behavior. Most robot authors subscribed to the popular notion—intrinsic to the Web's deepest structure—that putting data on an accessible Web server gave tacit permission for anyone or anything to come by and take a gander. Malfunctions were rare, argued robot authors, and usually helped point out weaknesses in Web site design that webmasters needed to address.

The robot author defense had its merits, except for one major fallacy. Web site log files demonstrated an increasing number of robot misdeeds. The system, in the opinion of both Ginsparg and Hartill, who monitored such logs, was collapsing.

The Web's own success was partly to blame. Population growth guaranteed an ever-greater number of novice robot authors testing out new robots. The increasing stresses on the Web's infrastructure caused by both human and robot overuse deepened the growing split between robot authors and webmasters. But that wasn't the only force at work in an increasingly dire situation.

Paul Ginsparg identified another culprit: greed.

"Everyone and his or her brother has decided to make a fortune on another Wall Street IPO with another 'indexing the internet' company," Ginsparg wrote in a message to the Web robot mailing list as the nastygram debate finally subsided.

Big business had arrived on the Net.

ROACHES AFTER DARK

As new, untested bots, heedless of etiquette and flawed in execution, jostled their way onto the Web over the course of 1996, one Web robot reigned supreme: Scooter, the leanest, hungriest bot in cyberspace.

Five thousand lines of tightly packed C code, Scooter is best of breed, smarter than Michael Mauldin's Lycos spider, faster than Brian Pinkerton's WebCrawler, more robust than Alex Cohen's Wobot. Scooter is wily—it knows how to skirt dangerous black holes and avoid nasty compressed files. And

Scooter is strong, backed by the industrial might of blazingly fast DEC Alpha 64 computers and a massively thick connection to the Internet. But most of all, Scooter is a speed daemon, able, in the early summer of 1996, to traverse the entire depth and breadth of the millions of documents on the Web in little more than a week.

Scooter is the bot behind AltaVista, Digital Equipment Corporation's hugely successful Web-based search engine. Throughout 1996, as hundreds of software companies flooded the Web with an endless succession of glossy, high-profile software products sure to be "the next big thing" on the Net, the tiny Scooter labored in the shadows, unbeknownst to most of the Web masses. But the little robot arguably did more to transform how the corporate world perceived the Web than any other entity had. Scooter changed the world. Scooter showed how bots could mint money.

"I like to tackle things I know to be impossible," says Scooter's creator, Louis Monier, AltaVista's senior consulting engineer. "And I like to build things that work." Scooter works.

Monier, French-born, barrel-chested, and bearded, refers to Scooter with affection and pride. Together, the man and the bot made AltaVista the number one search engine on the Web in an astonishingly short period of time, taking the Web by word-of-mouth storm. Of course, they didn't do it alone—Digital, Monier's employer, had more than enough resources to make Scooter fly. One floor beneath Monier's Palo Alto office, the computer lab that housed AltaVista made the point quite clear. Bots were no longer a hobby, a hacker's delight.

The lab is the kind of place geeks dream of: an info-nexus where ranks of squat, refrigerator-size machines seethe with potent silicon energy. Some of the machines have their front panels removed, revealing gigabytes of memory. Neatly strung cables loop across the ceiling. Thousands of tiny red lights flash on and off. Hard drives whir. Inside this room churns a universe of information. Questions are asked, and answered, at unfathomably fast rates.

AltaVista's database is an index that attempts to map every word on the Web. It is, as Monier notes, an impossible task— the Web is constantly changing, and much of its content is increasingly kept hidden from robot eyes, protected by password barriers and digital strings of barbwire defenses. But what remains still attracts AltaVista's attention. One day in July 1996, as Monier wanders through the lab, gesturing grandiloquently at the computers as if he had built them with his bare hands, thirteen million queries arrive at AltaVista. Thirteen million requests for the location of information—at the time, the total breaks the all-time record for a service officially only six months old. But a month later AltaVista records sixteen million queries a day. By the fall, twenty million. By January 1997, twenty-six million. The index itself contains the full text of thirty-one million documents and four million Usenet news posts (or articles, as they are sometimes called).

Granted, a hefty proportion of the blinking lights in Alta-Vista central signify requests for such worthy items as the location of naked pictures of *Baywatch* television star Pamela Anderson Lee. (WebCrawler's Brian Pinkerton once noted that fully 50 percent of the queries logged by his search engine were sex-related.) Still, all that digital inquisitiveness is impressive.

It is a public relations dream. AltaVista's success did wonders for Digital's public profile, and Digital execs have been made very happy by their investment in the search engine. Their contentment is underlined by two printouts of email messages from Robert Palmer, Digital's CEO, that are tacked to Monier's office door. Both laud Monier's above-and-beyond contributions to Digital's corporate welfare. Monier waves at them, smiling, but then jokes, "There is more broad recognition of the name *AltaVista* than the name *Digital*."

AltaVista's success marked a paradigm shift both for the Web's evolution and for the importance of bots in human affairs. No longer would the Web grow by happenstance. Digital's entry into the fray foretold the arrival of highly capitalized corporate heavyweights, a blue-chip roster of Dow Jones titans—Microsoft and IBM, Intel and Apple. By the fall, AltaVista had been spun off by Digital into its own stand-alone company, which sought ways to cash in on the search engine's ever growing popularity.

The profit incentive had come home to roost. Millions of queries—millions of hits—implied millions of eyeballs tracking their way through cyberspace. And where there were eyeballs, advertisers were soon to follow. As the Web penetrated mainstream society's consciousness, the business of information location, once an idle pastime for curious grad students, soon became recognized as having the biggest eyeball attraction potential of any Web-based enterprise.

Never before had a bot's performance meant so much in terms of cold, hard cash. Forget the chatterbots, gamebots, and IRC annoybots, the juvenile antics of botrunners and chat room imposters. As the Web hurtled its way from research network to commercial marketplace, Web robots led the charge. In a go-go market, where in a matter of mere months search engines and Internet directory sites blossomed into multimillion-dollar publicly held corporations, efficiencies in Web robot design had huge financial implications.

The ramifications of robot evolution did not stop at the profit-and-loss chart either. As Web robots pole-vaulted from university campuses to Wall Street brokerage houses, they announced a new era for the Net—politically, infrastructurally, and ethically. The transformation was as sudden as it was profound.

The robot authors who just a few months earlier were quaffing mugs of beer in Darmstadt became a budding bunch of millionaires in breathtakingly rapid fashion. Nothing of the sort ever happened to Eliza's Joseph Weizenbaum or Eggdrop's Robey Pointer. The high-profile success of robots like Louis Monier's Scooter had shattered the old bot way.

The changes came in quick succession. Brian Pinkerton signed a million-dollar deal to take WebCrawler to America Online. One of his first actions was to hire Martijn Koster, the former anti-robot avenger, to work on perfecting WebCrawler. At the same time, Michael Mauldin decided that running a start-up company was far more exciting than teaching at Carnegie Mellon. He secured funding by selling Lycos to the CMG Corporation, and he licensed the Lycos catalog rights to Microsoft.

Yahoo! also began to raise venture capital funding, setting the stage for a hiring spree and huge expansion. The McKinley Corporation followed suit. Yet another group of Stanford students formed their own company, excite, with backing from Silicon Valley's biggest venture capital firm and proceeded to rate Web sites in direct competition with Yahoo! and Magellan. An hour's drive north, at the University of California at Berkeley, a graduate student and a professor collaborated on a new approach to Web searching that employed parallel processing. They called it Inktomi (after yet another kind of spider) and formed a joint venture with *Wired* magazine's Web site, HotWired.

Other companies were quick to follow—the mighty and ponderous IBM even began to get its feet wet in the search engine business. The competition became ferocious. Lycos, Yahoo!, and excite all soon staged independent public offerings on the stock market. By the end of 1996, excite had swallowed two of its main competitors, acquiring both WebCrawler and Magellan.

The Web had changed, at its heart and core. One day, you're doodling at your computer, tinkering with some search algorithms, and the next, you're negotiating the terms of your stock option package. And suddenly, those search algorithms are trade secrets. Suddenly, Alex Cohen looks at his log files, sees the telltale track of a robot, and wonders if his competitors are trying to steal his data. Suddenly, the cooperative harmony engendered by the gift economy is dead.

A dramatic shift in the organizing principles of Web society followed in the wake of search engine mania. Where robot authors once traded tips on the best ways to solve certain problems or avoid certain disasters in the congenial, collegial atmosphere of mailing list "working groups" and academic conferences, now they patented their code compilations and kept quiet about new discoveries. The emergence of profit-and-loss considerations in the Web robot business led, on at least one level, to a positive disincentive to cooperate. The speediest robot ensured the best search engine—so why worry about overloading Web servers all over the Net? And why share your best algorithms with someone who might be looking to gain hits at your expense?

Consensus, it appeared, was the first victim of free market competition. The profit incentive expanded the hairline cracks in the Web programmer community into gaping chasms. As the battle for search engine and Internet directory eyeballs roared on, with larger and more voracious corporations piling into the fray, one had to look hard to find traces of the old Braustubl bonhomie. The era in which technonerds ruled the online empire was over—they no longer called the shots. Entrepreneurs had come out of the closet like roaches after dark. The flimsy fabric of the Robot Exclusion Protocol offered little protection.

PARASITES AND NEW PLAYERS

In this new era, Scooter, though a driving force behind AltaVista's surge to search engine supremacy, also happened to be one of the best-behaved Web robots on the Net. Louis Monier is a firm believer in the Robot Exclusion Protocol—he stresses its importance at every public opportunity. He is quick to respond to any criticism of Scooter expressed on public mailing lists and is always making changes to Scooter's code that uphold its reputation as a responsible Web citizen.

Monier's personal ethics played an obvious role in the maintenance of Scooter's moral character. But Digital's high corporate profile also provided a check on bot berserkness. Although the arrival of corporations onto the Web changed the game for everyone, corporations themselves had to be particularly careful about their actions. In some respects, they were more constrained, not less, than lone-wolf individuals. Corporations are financially liable according to their "deep pockets," and if one corporation discovered that another had infringed on its trade secrets or revenue-generating potential, lawsuits would be (and have been) immediately forthcoming. When Alex Cohen discovered the bot from his company's competitor rummaging through his Web server, he was alarmed, but he knew his defenses were strong. If the bot had actually broken through those defenses, a call to the McKinley Corporation's lawyers might have been the next step.

But even if potential corporate misdoings had limitations, the chain reaction set in motion by the introduction of fiduciary motives dramatically complicated Web robot ecology. And as the basic technology for building a Web robot spread into more and more hands, these new layers of complexity combined with the uncontrollable actions of thousands of individuals to utterly alter the Web's landscape.

Bot parasites began to appear—bots that piggybacked on the efforts of other bots. Some of these bots seemed innocuous, but every new iteration threatened another cascade of paradigm shifts.

The MetaCrawler bot, developed by Professor Oren Etzioni and a team of graduate students at the University of Washington, is one example of the new breed. The MetaCrawler is a Web robot that confines its search area to other search engines. In other words, if you input a query at the MetaCrawler Web site, MetaCrawler races over to the top search engines and directories—Yahoo!, WebCrawler, Lycos, AltaVista—and retrieves *their* results for that query. It then bundles them up in a nice, ordered package and delivers them to you.

A curious Web surfer might consider MetaCrawler to be a useful, timesaving tool. But the administrators of the original search engines had solid reasons to look askance at such banditry. A MetaCrawler user never saw the advertisements ostensibly supporting the original search engines. There was even the possibility that a MetaCrawler-style site might sell its own ads, thus reaping income from robotic work that by all rights belonged to someone else.

Even the most innocent robot trespass could have a direct negative economic impact on any Web site that depended on advertising for revenue. On the Web, advertising fees are usually pinned to the number of hits a site receives. If a rogue spider blunders into a site and crashes a Web server, creating a denial-of-service crisis, then the site starts to hemorrhage financially. Log file analysis saved advertisers from paying for robotic accesses, so one could even argue that every single time a robot occupied a server connection that a human could have been using, the Web suffered. The worst nightmare for commercial Web server administrators, of course, was the prospect of malicious denials of service—of competitors launching robots against their site with the full intention of knocking them out of business. In the early days of the Web, such attacks were rare—it was too easy to trace an offending robot back to its den. But as the Web became more complex, and programming tools like Sun Microsystems's Java and Microsoft's ActiveX became more prevalent and powerful, the possibilities for talented hackers grew steadily.

Each new advance in robot technology threatened the nascent financial relationships that had grown up around the last robot leapfrog. And experimentation was hardly confined to the aboveboard environment of a university or corporation. A whole shadowy underground of Web robot operators emerged from the murky depths of cyberspace. A new breed of Web robots, designed to work at cross-purposes to the standard Web indexing robots, staked out their territory on the Net.

Ivana, a sophisticated spamdexing bot, offered one of the most flamboyant examples of the new possibilities for Web robots. Developed and operated by Etoy, a troop of European artists and hackers who specialized in postmodern performance-art stunts, Ivana was a tool for hijacking unwary Web surfers into a disorienting Etoy Web lair dedicated to freeing the notorious criminal hacker Kevin Mitnick.

Ivana raised the art of spamdexing to new heights. First, Ivana's operators gave her a list of keywords to input into various search engines. Ivana then retrieved the top ten rankings for each word. For example, if the word was *Porsche,* she grabbed the ten Web pages that WebCrawler suggested were most likely to be relevant to Porsches.

Search engines determined relevancy via tightly guarded sequences of rules, or *heuristics*. If a document repeated a keyword a certain number of times, or if it appeared capitalized or in the title or italicized, the robot or indexing program had a better chance of guessing the document's relevance to a chosen keyword. Heuristically determined artificial intelligence is a state-of-the-art research area for search engine operators and a prime requirement for tomorrow's Web robots.

After retrieving the top ten Web documents for a particular keyword, Ivana then deconstructed the documents in an effort to discover what particular combination of words, placement, and formatting guaranteed such a high ranking. After unlocking the mystery—again, different for each search engine— Ivana and her masters would generate a fake document bound to rank up at the top of the list for any future seekers after a specific keyword. So, for example, an innocent Porsche fan might enter the word *Porsche* in the WebCrawler search engine and see an interesting-looking URL titled Porsches, Porsches, Porsches. But after clicking on that URL, the sports car enthusiast might suddenly be spirited off to Etoy headquarters, where the mischievous botrunners subjected all unlucky victims to garbled English rants about Mitnick (including a sound file of the Etoy team singing a Europop ditty in praise of Mitnick and their own search-engine spamdexing prowess).

The Etoy hackers were just one example of an ever-growing number of Web-wandering hackers intent on using Web robots to subvert established business models or push private cultural agendas. More appeared every day. Spamdexing wars, piggyback parasite bots, webmaster–robot author disputes—clearly, the Web was no longer one big happy family. Rogues, and rogue spiders, grew in number every day.

To an observer mindful of the example of the IRC botwars or the cancelbot frenzy breaking out all over Usenet, the only thing holding the Web back from unrestrained bot chaos seemed to be the technical proficiency required to program and operate a Web robot. But that state of affairs could not last. In the wake of search engine success and the ever-booming growth of the Web, market-driven commercial imperatives demanded that every product niche be exploited. And Web robots, which played on that irrepressible human desire to extend one's presence in cyberspace—that now familiar cyborg will-to-power—offered an inviting target. If the Web masses couldn't build their own bots, then somebody else would. In the new, postcapitalist Net, anyone could be a bot-powered rogue.

SPAMBOTS ON THE LOOSE

One of the first signs that bots were poised to break out of their techno-ghetto came in May 1996, when GlobalMedia Design, a Florida-based Web site production house, announced the release of RoverBot. A commercially marketed robot, RoverBot—or Rover, as the bot is also called—represented a giant leap forward in the laborious trek from prehistoric daemon to near-future intelligent agent.

The Rover Search Service did not actually hawk the robot itself, but rather the product of its labor. RoverBot's Web page bragged about "a unique Internet research tool that generates custom email mailing lists by exploring web pages that meet your criteria." "Your criteria" meant keywords for Rover to chew on. Rover inserted those keywords into a few popular search engines, generating lists of Web documents theoretically relevant to the search query, and then retrieved those documents and examined them for email addresses. (Technically, Rover looked for a particular kind of HTML code called a mailto button that indicates a "clickable" email address—click on the button and a simple mail program sends a message to the mailto-button addressee.) Those email addresses were then compiled into a list for the RoverBot customer.

Rover exhibited no dramatic advances in Web robot technology. The breakthrough was to market Rover's services, for a fee, to the masses. In essence, GlobalMedia Design was selling the ability to construct lists most suitable for direct-mail advertisers. Rover was a spambot—perhaps the most hated genus of bots in the entire bot catalog.

Robots excel at generating spam, a term prevalent on the Net that describes the electronic equivalent of junk mail—unsolicited messages of a commercial, political, or otherwise nonpersonal nature. Rover itself, however, did not create spam—that task fell to the people who bought Rover's mailing lists. Rover was actually a quite decent bot. It did not ceaselessly search the entire Web, gnawing away at bandwidth and poking its snout where unwanted. Brian Clark, the president of GlobalMedia Design, considered himself a part of the "old Internet" and made sure that Rover scrupulously followed the Robot Exclusion Protocol. Rover's Web page included paeans to netiquette and even advanced the Robot Exclusion Protocol by allowing wary webmasters to exclude Rover by inserting a hidden HTML tag known as the "baddog" tag into their Web pages. Rover's programming required that the bot, when retrieving a page, look first for that tag; if it found the baddog notice, it had to drop the page and skedaddle.

Possible inclusion on automatically generated junk mailing lists is not usually why a webmaster or anyone else includes an email address on a Web page. But such spam is a fact of online life. A single contribution to a popular mailing list, or a one-line post to a Usenet newsgroup, will attract the attention of hordes of Rover-style bots, hard at work generating lists for bottom-of-the-barrel advertisers looking for the cheapest possible way to get their message out.

Brian Clark once toyed with the idea of creating a companion to RoverBot that he jokingly referred to as the HappyBot. The HappyBot would have been a kind of Barney bot of the Web, wandering around committing senseless acts of kindness in every mailbox it found. But Clark refrained. He had a good sense for what the old guard of the Net considered acceptable bot manners. An automated "Have a nice day" bot crossed the line.

So RoverBot followed the rules. But that did not lessen the threat it represented. RoverBot's commercial release foreshadowed the spread of a malign new growth on the great Tree of Bots—bot proliferation into the hands of the Web masses. Once bots became commercially available, user-friendly, and accessible to anyone with a connection to the Net and a few dollars to spend, the ability of the Net's ancien régime to maintain the status quo would be severely weakened. Brian Clark was just one man. Others might not be so restrained.

THE FLOODGATES OPEN

One morning, going through my email, deleting the normal day's load of spam and mailing list detritus, I stopped and stared at a mysterious message. The sender of the message identified itself only as "HipCrime." The title of the message, likewise, was "HipCrime." And the body of the message contained only a single line of text—the URL *www.hipcrime.com.*

URLs arrived in my mailbox all the time, but usually with some kind of description that allowed me to easily ignore them. This one piqued my curiosity just enough for me to fire up my Web browser and investigate what turned out to be a run-of-the-mill Web site attempting to pass itself off as an "Internet Art Work."

About a week later, I received a second missive from HipCrime. Again, the URL, followed this time by another promotion for *www.hipcrime.com*, along with a message the gist of which ran "I really like your Web page and invite you to look at mine."

Humph. I had no Web page, though my email address, I knew, was widely available. A spambot had me, dead in its sights. I configured my mailbot to reject any further mail from HipCrime, but I was aware, even as I did so, that I was fighting a losing battle. Like pests that evolve immunities to each new onslaught of deadly insecticide, new spambots, more powerful, wily, and obstinate than their predecessors, mutate in response to each new obstacle placed in their way. And the annoyances they cause are accidental no longer, a fact that became crystal clear several months later, when the author of the HipCrime spambot revealed himself to the world.

Signing his messages "Robert Returned," he appeared simultaneously on several mailing lists devoted to agents and robots. To an audience flabbergasted at his audacity, he declared that the HipCrime promotional messages had been generated by a Java applet he named ActiveAgent.

ActiveAgent broke every rule in the Web robot book. It ignored the Robot Exclusion Protocol. It scoffed at antispam netiquette and sneered at horrified expressions on webmaster faces. Starting with any given URL, it followed every link it could locate, wherever it was led, throughout the entire World Wide Web. Like RoverBot, it searched for clickable email addresses. But unlike RoverBot, ActiveAgent made no attempt to distinguish whether the email addresses it found were relevant to any particular topic or predetermined keyword. Instead, when it found such an address, it pushed the mailto button and sent a message.

On its initial trial run, the message sent to thousands of people contained only the URL for Returned's Web site: *www.hipcrime.com*. Simple but effective. Returned knew that just getting his URL into as many mailboxes as possible would encourage recipients to do just as I had done—to come take a look at HipCrime. ActiveAgent served as Returned's robotic, roving publicity agent, tirelessly seeding the Web with mini press releases. And it worked, according to Returned, who claimed that more than a thousand visitors had hit the Hip-Crime Web site within the first eight hours after Active-Agent started operating.

It did not bother Returned in the least that most people condemned unsolicited mail as a sin worthy of a new Commandment, or that they considered spam one of the foremost evils of online life. Returned welcomed the negative attention spawned by ActiveAgent. Any publicity equaled good publicity. The more people who complained in public forums like mailing lists and newsgroups, the more people who heard about his Web site. He even went out of his way to taunt protesters, thanking his "detractors" for "making so much noise that traffic will remain high for a long, long time."

He exulted in the cries of wrath. "The raving, angry notes can be a source of great enjoyment," he wrote in a post to Koster's Web robot mailing list.

Returned delighted in thumbing his nose at the responsible robot author and webmaster communities. In contrast to Brian Clark, who proactively facilitated webmaster efforts to block RoverBot, Returned rejected all requests that he provide methods for thwarting ActiveAgent. ActiveAgent's goal, he stated, was "to locate as many email addresses as possible." Anything that interfered with the "completion of this assignment" had to be considered a bug, not a feature.

Returned's reckless demeanor was bound to ruffle feathers and raise hackles all over cyberspace. But the kicker to the Active-Agent story had less to do with Returned's attitude than with his marketing policy. Returned had big plans for ActiveAgent. Just as chatterbot author Ken Schweller planted his user-programmable bots in every MUD he could find, Returned seemed dead set on propagating his ActiveAgent bot as widely as possible. For a fee of $100, Returned sold ActiveAgent's source code to all comers. ActiveAgent purchasers could then turn around and program the bot to send whatever message, however objectionable, they desired. The possible consequences of encouraging spambot overpopulation bothered Returned not one whit.

Returned's style fit the profile of the average member of an IRC botrunning gang. But in the context of the typical discourse to be heard on the technically oriented mailing lists, his insouciance struck an off-key chord. Even during the worst flamewars, most robot authors and webmasters expressed concern about the implications of their actions for the general welfare of the Web. But not Returned. Like an arms dealer happy to sell weapons to both sides in a bloody conflict, Returned felt no compunctions.

< >

Returned's attitude may have been redolent of juvenile obstreperousness, but his actions banged the gong for the future of bots in cyberspace. The widespread distribution of the ActiveAgent bot presaged the arrival of a new and troubling world, one in which powerful tools are within the grasp of every user, in which human-computer cyborgs exercise power in an arena of unfettered anarchy. In the post-ActiveAgent future, a phalanx of MeanAndNastyBots will follow close behind every innocent HappyBot. Millions of bot-empowered Web inhabitants will rush to test out their new superhuman abilities, unconstrained by the flesh-and-blood restrictions on human interaction that prevail in offline life.

The first signs of that transformation are already written on the wall. It is a trivial matter to adjust the ActiveAgent bot so it endlessly crisscrosses the Web, blitzing whomever the botrunner wants to harass with hatemail or mailbombs. The possibilities range from the tiresome—one morning, say, everyone on the Web who has an email address on a Web page containing the word *China* receives a virulent anti-Asian screed in their email box—to the troublesome. Communist China, right now, blocks Internet access to certain Web sites that promote "dissident" views or alternative versions of approved Chinese history. What is to stop a Chinese patriot from taking a more active stance, from commanding a bot to bombard a Web site devoted to Taiwanese independence or Tibetan human rights until it chokes to death?

Of course, mailbombs and spambots do not mean the end of the world. Real bombs pack just a bit more punch. The bot hostility represented by spambots, denial-of-service attacks, nastygrams, and spamdexing hoaxes are tame gestures in comparison with real-world violence and social misery. Junk mail can always be deleted, and an inaccessible Web site rarely spells permanent disaster, even if it does impact advertising revenue.

But the ActiveAgent spambot is the first trickle in a flood. The spambot surge is significant, not because junk mail entails the collapse of civilization but because it reveals how bot use will always reflect a Manichean duality. Evil bots will accompany good bots—so it has been from the beginning and so it will be until the end. Bots may have originally evolved as playthings, as experiments in artificial intelligence, or as solutions to problems arising in complex computer-created environments. But bots are far from limited to playing only those roles. As the interface between human and computer, as anthropomorphic reflections of our own deepest character traits, as plowshares that can be hammered into swords, bots will exemplify all our deepest hopes *and* fears.

The narrative of bots on the Web has hardly begun, but already the story line is clear. As the potential applications of bots and agents increase, so does their accessibility—bot proliferation is an out-of-control arms race in which everyone on the Net packs a pistol. We have become trapped in a vicious circle. As the Net increases in complexity, we require ever more powerful tools and helpers. But as our tools grow in potency, their potential misuse becomes ever more devastating, forcing us in turn to gird ourselves with ever stronger defenses.

The first bots have hardly pulled themselves onto dry land and begun climbing up the evolutionary ladder. But the fact that they are single-celled weaklings compared with the complex organisms following in their wake should be as much a warning to us as a comfort. We might as well assume that as bots become more powerful, they will also be employed in the pursuit of nefarious ends and will run spectacularly amok. As the scale of cyberspace existence expands, so also will bot mayhem.

< >

The implications of bot duality become even more serious when we stop to consider the visions that a world of researchers and entrepreneurs have for botlike programs. Bots, especially Web robots, are the first rough-and-ready stabs at real intelligent agent implementations. And the rhetoric of agent potential respects no limitations.

Nicholas Negroponte imagines personified agents taking care of all our most pressing needs—to the point that they are driving our cars and making our beds. Microsoft's Bill Gates dreams of agents that control every aspect of a person's home—air-conditioning, alarm clocks, coffeemakers, as well as computers. Even if the fulsome detail of their respective visions comes across as somewhat loopy, they are not entirely off base. There is little question that both the Net and the Net's indigenous creatures will become more tightly interwoven into the fabric of our daily lives. Microsoft alone, pouring billions of dollars into a research and development division that includes some of the world's top scientists in the fields of natural language processing, computer graphics, and agent technology, will relentlessly push the integration of personified autonomous semi-intelligent programs into every facet of our consumer lifestyle. And Microsoft, as mighty as it is, just reflects the even more awesome trends of life in capitalist reality. The marketplace demands new products. Agents, and bots, will be rushed to fill that demand.

The primitive Web robots scuttling from hyperlink to hyperlink today represent only the most shadowy inklings of what is to come. They are simple, rudimentary programs that have seized popular attention partly because the concept of an autonomous bot is inherently sexy and frightening. But their capabilities are as dust before the avalanche to come. Thousands of researchers all over the world are hard at work constructing new protocols and languages that will allow autonomous programs to communicate with each other, make decisions, react to environmental changes, and carry out tasks. The Net will soon experience the debut of "distributed agents":

swarms of bots acting in concert, updating each other and correlating events all over the Net. These agents are intended to run air traffic control systems, to monitor the flows of capital through financial networks, and to keep factories and telecommunication nodes operating smoothly.

Progress is incremental but steady. And it is vitally important. The Net, as an entity, is already beyond any one human's cognitive grasp. It will only become more so. We will be lost without our bots.

But as these agent/bot armies assemble in the offing, we should already be preparing for bot-driven denial-of-service attacks against those very factory-agent systems and autonomous traffic-control networks. Clogged bandwidth will be the least of our problems. Once autonomous programs are at work in every sector of the networked future, real lives will be at stake when those programs accidentally misfire or are maliciously abused. And the more autonomy that is injected into the Net, the less control any individual entity—human, corporate, governmental, or digital—will have.

The Web is rife with examples of bots twisted in purpose. Within a year of their creation, Web search robots doubled as stealth bomber spybots, as Alex Cohen discovered. The BargainFinder CD shopping bot can be used to prevent human shoppers from gaining access to an online store just as easily as it can be employed to get information out of the store. There is now a bot that will watch the Web-connected Federal Express package database and alert you when your personal package has been delivered. But any bot that watches a site for particular occurrences is also well suited to be a surveillance bot, tracking human movements through cyberspace.

And therein lies one of the most troubling possibilities of the bot future—the employment of bots to gather information and create dossiers about individuals. Brian Clark's RoverBot gathered such information on an extremely simplistic level, searching the Web for matches to a keyword. But what if it was searching for matches to the word *sex,* and once, late at night, you had made the mistake of signing the guest book at Bianca's Smut Shack Web site in a fit of inebriated passion. Now, forevermore, your email address is associated with a sex cross-reference, and suddenly you are considered fair game for every porn merchant trolling the Web.

Bot-created profiles won't be confined to lists of email addresses. As bots become smarter, and the tentacles of the Net intertwine themselves further and further into the minutiae of our existence, bots will begin constructing profiles of human behavior that dwarf what is currently possible from reviewing a credit card trail or a barcode-generated record of supermarket purchases. As online transactions grow to dominate economic life, and databases of information of all kinds are increasingly brought online, the imprint of our personal treks through life will be more accessible to bot fingertips.

RoverBot, the Next Generation, may comb cyberspace to generate a profile of you for a marketer, a police officer, or a terrorist. A citizen applying for health insurance or a mortgage loan may find that before approval the officer at Mutual of Omaha or Bank of America sends a couple of bots into cyberspace to dig up all the dirt on that citizen they can find: the posts to alt.sex.fetish.xxx, the humorous references to recreational drug use on a homepage, the participation in a mailing list discussing methods for blocking bad credit reports. A visit to the Smut Shack—boom, your petition to adopt a child is rejected. And as the Net becomes an ever greater forum for our cultural and community interactions, the potential for devastating juxtapositions becomes more pernicious. What happens when a Microsoft bot correlates some questions you asked in a newsgroup about Microsoft's Word 97 program with the fact

that it has no record of you as a registered owner of that program? The Big Bot Brother will always be watching.

Alex Cohen's discovery of a bot searching through his Web
server made him contemplate the potential of bot use for commercial skullduggery. But he was able to pinpoint the intruder
as a bot without much ado and could have painlessly blocked
it should he have so desired. But defending against bot and
agent incursions is going to get much harder. Cohen's fellow
researchers, so hard at work concocting elaborate new frame-
works for agent technology, will see to that. Bots and agents
are being designed that will purposely mislead humans and
other bots about their intentions.

Current bot mischievousness will not compare to the onslaught of reality warping soon to be engendered by successive
waves of these new "self-interested agents." The term refers to
agents out to get the best deal for themselves (or their masters). Agents that are greedy, that only share information when
they have to, that know when to lie and when to cheat.

For example, if your agent is shopping for a used computer
from an agent with computers to sell, it is not in either agent's
best interest to play it straight, to indicate how much money it
has to spend or what the bottom-line sales price is. Negotia-
tion often requires deceit. So in addition to spybots and surveillance bots and spambots, we will have bots that say one
thing and mean another, that disguise themselves to get past
checkpoints, that manipulate other bots and humans for their
own advantage. Remember that search robot looking for information on sexual proclivities? You'll never even know it is
there, as it cruises the Net pretending to be AltaVista's Scooter.
That chat room participant singing the praises of Apple's new
operating system? He's a lying, no-good propagandabot.

It gets worse. On IRC, unscrupulous channel protection
bot writers have been known to include "back doors" in their

programs that allow them to sneak in and take over a channel or to steal login passwords and tamper with online files. Defending against commercially sold or shareware Web robots that come riddled with back doors or other unknown traps will become a staple of life on the Net.

Individuals will always harass individuals. But corporations will also be battling with other corporations in cyberspace. It's not a new phenomenon. Industrial espionage and fierce struggles over intellectual property have long been staples of corporate life. Most corporations are currently suspicious enough of the Net that they protect themselves from it by "firewalls" separating internal corporate "intranets" from the wild Net. It doesn't require an overfervid imagination to conclude that someone, somewhere, is working on bots designed to sneak through firewalls and grab useful data. Bots are creatures of information—they will flourish in an era when protecting or accumulating information is increasingly the benchmark activity of economic life.

Computer scientists have a host of responses to alarmist warnings concerning potential bot havoc. Cryptographic authentication, "safe" network transmission protocols, legal defenses, and a myriad of other fixes will be applied as bot prophylactics. Some of them may work; some of them may not. Their efficacy isn't the point. Their necessity is the issue. Bots and agents will be abused, both by corporations and governments invading our privacy and promoting their own agendas and by millions of individuals eager to kick sand at each other in the online playground.

Such scenarios put the bad bot realities of IRC and MUDs into illuminating context. The Web is rapidly becoming the platform in which all forms of information, private and public, merge into an interconnected whole. We can live a long life without ever needing to enter an IRC chat room or pretend in a MUD to be a battle-ax-wielding dwarf, but we can't avoid the tendrils of the Web. We can't avoid the curiosity of a Web robot.

As in MUDs, IRC, and Usenet, bots on the Web are more than autonomous helpers—they are prosthetic extensions of the human body in cyberspace, a means for exerting power in the digital realm. The exertion of such power almost always causes friction. In IRC such friction nearly destroyed the network infrastructure. How much greater levels of stress will be caused by the combined action of millions of Web users, all armed with bots and agents, all exercising their ids at play in the fields of cyberspace? It may be easy to dismiss the botwars of IRC as driven by teenagers working out their hormonal angst. But how will the greater Net respond to pressure from hustlers motivated by commercial gain, bigots driven by ideological hatred, and multinational corporations with a global reach and unlimited bank accounts, intent on maximizing their bottom line?

The Web represents the full flowering of the decentralized anarchy so fundamental to the Net's evolution. The real world has its rulers, its courts of law, its federal inspectors, police, and jails. The society of the Net shies away from those entanglements. But in the absence of outside restrictions, the Net's own citizens are scorning self-generated guidelines like the Robot Exclusion Protocol. They are rushing headlong into an era when bot posses flank every move a Net user—man, woman, child, or corporation—makes. Will the bot warfare to come lead to a bot blacklist, as it did on IRC? Or will the Net offer another way?

7

ON THE BRINK

USENET IS BROKEN

Striking without warning, a cancelbot massacre hits Usenet on September 22, 1996. It eliminates nearly thirty thousand messages. Every single article posted to entire hierarchies of newsgroups vanishes overnight, brought down by a barrage of carefully tailored software missiles. Usenet's normal raucous hubbub aborts; its torrent of clamorous debate on everything from the latest Chow Yun Fat performance in alt.asianmovies to 3-D rendering bugs in comp.lang.vrml is suddenly stifled.

A band of battle-scarred Usenet veterans hanging loose in news.admin.net-abuse.misc, Usenet's central clearinghouse for spam alerts and network abuse reports, springs into action. As details of cancelbot atrocities pour into n.a.n-a.m from every Usenet sector, the veterans launch countermeasures. Chris Lewis, a master bot operator and one of Usenet's foremost cancelers of spam, activates his Dave the Resurrector bot, a kind of anti-cancelbot that specializes in reviving improperly canceled articles. Others work on alerting the administrators of the major Usenet news gateways to the crisis. Their efforts are soon rewarded. Almost as quickly as it starts, the frenzy ends.

The next morning, thousands of regular Usenet readers demand to know who canceled their postings and why. Many of these readers have never before paid attention to cancelbots. Now the word is on everyone's lips. The calamity is so overwhelming that even the *Wall Street Journal* takes notice. The paper's coverage quotes one software engineer's summation: the attack is "Internet terrorism."

The denizens of Usenet are unaccustomed to such a high profile. But it does them little good. Their outrage is vented in vain. Few answers are forthcoming in the wake of the massacre. No one knows for sure why someone would want to squelch Usenet, a citadel of free speech. There isn't any rhyme or reason to the pattern of newsgroups chosen for such slash-and-burn tactics. Among the forums targeted by the cancelbots are those devoted to topics as diverse as pornography, the Middle East, and computer operating systems. The origin of the attack is traced to an account at a small Internet service provider in Tulsa, Oklahoma, but the trail stops there—the owner of the account paid in advance, in cash. The miscreant remains at large.

In n.a.n-a.m the prevailing view is that the identity of the "mass cancel incident" culprit can be found on a short list of names. Although cancelbot operation is not considered a challenging programming feat, these bots betray enough familiarity with certain technological details to suggest that their mastermind is someone well known to Usenet's core group of "approved" cancelbot operators.

Clearly, the attack is malicious and premeditated. Normal spam-canceling bots identify themselves to Usenet news servers with a title describing their purpose. A cancelbot targeted at the long-running Make Money Fast financial scam is labeled "Make Money Fast." Cancelbots meant to prevent large picture files or software programs in binary code format from being posted to text-only newsgroups are tagged as "bincancel-bots." But the mass cancel cancelbots bear names calculated to be obnoxious, rather than informative. "Kikecancel" hits news-groups devoted to Jewish culture. "Geekcancel" goes after computer newsgroups. "Slanteyecancel" spikes Asian news-groups.

The n.a.n-a.m regulars mull over the names of the usual suspects, a motley crew of Net-Kooks who can be counted upon for periodic outbursts of irrational behavior. But there is no compelling reason to believe that any particular person is guilty. What is the motive? Did someone, somewhere, say something that upset the cancelbot operator so much he or she decided to raze Usenet to the ground? Or is there a more dia-bolical subtext?

One theory hatched in n.a.n-a.m that quickly gains adher-ents hinges on the suspicion that the cancelbot operator is ac-tually a spammer—someone who has been showering Usenet newsgroups with annoying advertisements. By launching such a mass attack, the argument goes, the spammer is acting as a bot provocateur, aiming to bring all cancelbots into disrepute. The ultimate goal of such a campaign would be to force Usenet news server administrators to ban cancelbots, thus leaving the field free for the spammers.

The mystery of the cancelbot runner's identity is never solved—a not unusual case of affairs on Usenet, where ano-nymity is a cheap commodity. But in the long run, questions of who or why are immaterial to the real problem facing Usenet. This is not the first time automated malevolence has besieged the network. In fact, in the months prior to this mass cancel attack, there has been one outbreak after another.

Memories are still fresh of the "ellisd" incident, a cancelbot outburst similar to the mass cancel attack in every aspect except scale, and the "poetry festival," a spate of thousands of articles containing incomprehensible computer-generated gibberish. Nor has anyone forgotten "the pig bomb dropped on n.a.n-a.m": a fountain of Usenet posts that cascaded through the network and provided a defining example of uncontrollable Usenet "spew." Most recently, the ARSBOMB, another spew grenade, had been lobbed into the alt.religion.scientology newsgroup, where fans and foes of the Church of Scientology regularly assemble for chaotic and acrimonious debate.

Consensus is never easy to reach on Usenet, but for Chris Lewis and his n.a.n-a.m cohorts, the picture is sharp. Usenet is broken. After a summer in which spam, spew, and mailbomb mayhem have slammed into Usenet like a season of typhoons battering against the southeastern coast of China, this particular cancelbot attack may be the final straw.

Usenet needs help.

"Do we cave in to terrorism," writes Lewis, as the cancelbot floodwaters recede from n.a.n-a.m, "or stand firm and try to save Usenet from turning into a useless morass of gibberish?"

"DATA-DRIVEN"

> Any node can submit an article, which will in due course propagate to all nodes. A "news" program has been designed which can perform this service. The first articles will probably concern bug fixes, trouble reports, and general cries for help.
>
> —TOM TRUSCOTT, *cofounder of Usenet,*
> *at the 1980 Usenix conference*

The computer science graduate students at Duke University and the University of North Carolina who created Usenet in 1979 imagined their news service as a means for sharing information about that geekiest of all technogeek pursuits—Unix programming. They never dreamed it would one day turn into the biggest electronic bulletin board system ever devised, a forum for international communication that virtual community specialist Howard Rheingold likes to call "the world's largest conversation."

Unix is a computer operating system developed at Bell Laboratories and much beloved by programmers with get-beneath-the-hood bents. It's also an unruly mess. Numerous official and unofficial versions of Unix exist simultaneously, a smorgasbord of multiple "flavors" that vary according to the idiosyncrasies of major computer corporations and the volunteer efforts of obsessed hackers. Mastering Unix means keeping abreast of a never-ending flow of upgrades and patches and add-ons. It's a job too big for any one human.

But not for the parallel-processing power of the Net's collective mind. Or at least, that's what Tom Truscott and his colleagues hoped. If they could tap the combined knowledge of the international Unix guru community, no question would remain unanswerable for long.

They ended up hacking together an ingenious system that allowed computers to connect to each other over regular telephone lines and exchange articles. They chose the name *Usenet* for this new information sharing system, in a nod to the annual Usenix ("use Unix") conferences that had long been one of the major forums for exchanging cutting-edge Unix information.

Usenet eventually became tightly integrated with the evolving Internet infrastructure, leaving the era of direct dial-up connections between computers behind, but the underlying principle of Usenet news transmission remains the same today as first conceived in 1979. Each computer running the Usenet news transfer protocol is a network *node* or *news server.* A steady stream of articles—the Usenet *news feed*—flows from node to node, fed by thousands of tributary streams from each individual node.

That node-to-node style of information propagation, unhindered by central control and endlessly reconfigurable, offers a potent metaphor for understanding how the Net works, in terms of infrastructure, culture, and politics. Online communities like IRC, Usenet, and the Web grow by accumulating more nodes, and the complexity of each environment is furthered by the rapidly multiplying interconnections that stretch like an infinite latticework among all the multitudinous points of information. Each node is independent but must be able to communicate effectively with all the others in order to flourish.

Throughout the 1980s the infrastructure to support Usenet spread from campus to campus and corporation to corporation in the same node-to-node fashion as did the information it transmitted. Usenet, like IRC, is a completely decentralized product of the gift economy ethos that is also a self-contained unit, with its own customs, atmosphere, and ideology. But it is much, much bigger than IRC. Today the original focus on Unix has been left far behind, and millions of people regularly read and contribute to Usenet. (In July 1996, for example, Usenet distributed nearly 170,000 messages a day among 19,000 separate newsgroups devoted to every possible topic.)

In an information environment as large and fertile as Usenet, it doesn't take long before botlike creatures arrive, seeking sustainable niches in the new info-economy. A few, very rare bots posing as humans infiltrate some newsgroups—an AI postingbot like Mark V. Shaney, for example. But for whatever reason, they don't thrive. Usenet, like the Web, is a mechanism for managing the transfer and intersection of vast amounts of data, more information than any single person can administer. In such a scenario, helpbots are far more vital to the ecology than entertainer bots. Usenet's first proto-bots were maintenance tools necessary to keep Usenet running smoothly. They were cyborg extensions for human administrators who could not single-handedly cope with the immensity of networked existence.

Usenet's rapid growth required autonomously acting assistants that could perform tasks in a decentralized and unmappable environment. The first assistants to appear owed their genesis to mundane outbreaks of Usenet confusion. Originally, Usenet consisted of a flat hierarchy of groups with the prefix *net*—such as net.singles and net.suicide, Mark V. Shaney's favorite hangouts. But such an organizational chart proved inadequate to the multiplying array of subjects that Usenet readers felt compelled to comment on. In 1986–1987, in a reorganization grandiloquently referred to as the Great Renaming—as if it were a Vatican synod or constitutional convention—Usenet's existing newsgroups were regrouped into new hierarchies with descriptive names: biz., rec., soc., news., sci., comp., talk., and alt.

The renaming, though necessary, had some unfortunate consequences. Usenet readers unaware of the hierarchy changes continued to post articles to newsgroups now no longer in existence. One of the leaders of the Great Renaming, Gene Spafford, a computer science professor at Purdue University and the closest thing to an archwizard in the history of Usenet, came up with an off-the-cuff solution. He wrote a series of programs that scanned the Usenet news feed at selected choke points. These programs examined the identifying header of each article passing through the bottleneck, searching for incorrectly addressed articles and redirecting those it found.

Spafford describes his Usenet maintenance programs as

"early examples of data-driven programs that interacted with the network." But he demurs at calling them bots. Spafford is stingy with names and reluctant to personify or anthropomorphize computer programs, believing that to do so often leads to an inappropriate distancing of oneself from one's tools. The connection between tool use and tool user should be kept as tight as possible, argues Spafford. The looser that connection becomes, the more likely that tool users will act irresponsibly.

But whether one calls them data-driven or daemonic, the programs Spafford and several of his colleagues simultaneously released onto Usenet in the late eighties were the first true examples of botlike programs to inhabit the newsgroup network. They were prime examples of programs intended to extend human power in cyberspace. Stationed at nodal gateways, they filtered the Usenet news feed according to the parameters set by their masters.

And as they interacted with the network, they positioned themselves as the direct ancestors to the cancelbots soon to arrive (although well after Spafford had given up attempting to steer Usenet in productive directions). Cancelbots, too, pore over the news feed, examining message headers for specified criteria. Cancelbots, too, are the digital manifestation of human desire in an online reality. There is a crucial difference, though: if a cancelbot finds what it looks for, it does not simply redirect a message but instead consigns it to the Usenet dustbin. Cancelbots are one step beyond maintenance tools—they're police officers, hit men, and online watchdogs. And like any tool of authority, they must be wielded with caution.

"The main thing about cancelbots is that they are dangerous," says Homer Smith, a Usenet bot writer of some renown. "They can cause a nuclear meltdown. Since they respond automatically to postings, if they respond to a spam, they can spam the Net right back. And if they start to respond to their own spam, then you can see there won't be much of civilization left when you get up from bed eight hours later to face the mess you have made."

THE ANTI-PATHOGEN

C. What is a cancelbot?

A cancelbot is a program that searches for messages matching a certain pattern and sends out cancels for them; it's basically an automated cancel program, run by a human operator.

D. Sounds cool. Where do I get one?

If you have to ask, you don't get one.
—*The Cancel FAQ*, maintained by Tim Skirvin

When Dick Depew first started reading Usenet newsgroups in the early 1980s, he remembers, "the whole feed could be read in an hour or so." Just as World Wide Web Wanderer author Matthew Gray could at one time personally visit all the sites on the Web, so too could Depew single-handedly master the full breadth of the Usenet experience.

Even so, it took years before Depew screwed up his courage to actually start posting. Like most Net newbies, he lurked in his favorite newsgroups for quite some time before finally plunging in. He read technical groups and attended to bug reports for his favorite software programs, but otherwise kept mum.

Then, one day in the late eighties, Depew, an avid bird-watcher, posted an article to the newsgroup rec.birds. He never looked back. Not only did he become a profuse and vol-uble contributor to a wide range of groups, but by 1990 he had also taken steps to elevate his status far beyond that of the normal Usenet conversationalist. He joined the Usenet technocracy. He set up his own news server to act as a Usenet node in Akron, Ohio, where he taught immunology and bacterial genetics for the Northeastern Ohio Universities College of Medicine. As is true in IRC, server administrators—the owners or operators of the machines that serve as network nodes—are the de facto ruling class. They have no authority over machines other than their own, but at their server they can execute software commands that directly impact the Usenet news feed, or they can permit the operation of locally stationed bots for the same purpose.

As a news administrator, Depew's interest in the health and maintenance of Usenet became more than incidental. He began to view the overall quality of Usenet discourse as a problem that he could help solve. The flamewars and constant outbreaks of incivility endemic to Usenet alarmed him, as they did many online observers. He was convinced Usenet was headed for a crisis. As the conferencing system left behind its research-community roots, becoming steadily more populated and diverse, its utility as a forum for reasoned debate and convivial conversation had embarked on a long downcline.

Depew knew what to blame: anonymity.

One of the most controversial aspects of life in cyberspace, anonymity is a curse—or a blessing, depending on whom you're talking to—in every online community, whether IRC, MUDs, Usenet, or the all-encompassing Web. By reducing physical presence to packets of data, the Net veils clues to true identity. Whether one is expressing oneself in ASCII text or as a 3-D avatar, one is still hidden in an impenetrable mist of words and adopted images.

Depew hated anonymity. He believed that irresponsible individuals took advantage of the shelter of anonymity to gain cover for the worst kind of childishness and spiteful behavior. Usenet's easy anonymity, he believed, would doom it.

"Anonymity in general discussions is out of place," says Depew. "It does not contribute to the sense that people are responsible for their own words."

Within the lax security of the Net, setting up a fake account from which to post Usenet articles is a trivial matter. But in Depew's view, the really irresponsible people were those who set up services designed to encourage people to be anonymous—the operators of so-called anonymous remailers. An anonymous remailer, like a news server, is a computer that receives electronic messages and then forwards them to a new destination. But in the process of doing so, it strips the message of all normal identifying information, offering a cloak of invisibility popular to a wide range of people, from political dissidents to porn distributors.

In 1993, anon.penet.fi, owned and operated by a Finn named Julf Helsingius, was by far the most famous anonymous remailer. Depew reserved a special antipathy for Helsingius. He regarded Helsingius as a symbol of everything that was wrong with Usenet, as an irresponsible netizen with no thought for how his actions affected the greater community.

In the early months of 1993, Depew feuded openly with Helsingius, demanding that he shut down his remailer. Helsingius refused. Then, in mid-March, Depew announced to the newsgroup news.admin.policy that he was about to test a program he called ARMM—automated retroactive minimal moderation.

"I am testing a shell script to carry out 'Automated Retroactive Minimal Moderation,'" wrote Depew, defending his program as a response to Helsingius's own suggestion that moderation was the only way to control anonymous posting to groups that don't want it. "It cancels articles posted from anon.penet.fi."

Depew's biology background had inspired him to concoct a medical solution. He created an antidote—or, more appropriately, a toxin—to kill the anonymity virus. Depew's antidote, ARMM, was a cancelbot.

A cancelbot takes advantage of one of Usenet's built-in features—the capability that allows Usenet posters to take back, or cancel, an already posted message. With most newsreader software, all that is required to cancel a message is a single key press or mouse click. Hit Cancel and the newsreader launches the cancel message—a special kind of Usenet news article that travels across Usenet, instructing news servers to remove the original message from their news spools.

Anyone who has ever sent an email message in a fit of indignation or fury has learned how online existence facilitates rash action. A cancel function is a very useful tool in such circumstances. But the Usenet cancel function included a gaping loophole. Given the idiosyncrasies of the Usenet news software, only a trifling programming effort is required to forge a cancel message for any Usenet article, whether or not one is the original author of the message. So anyone conceiving a hatred for Dick Depew, for example, could painstakingly forge cancel messages for every article he or she saw that had been posted by Depew. Or if discussions of Armenian genocide drove a reader batty, he or she could spend days canceling all messages that even hinted at the topic.

Of course, if Depew is posting ten or twenty messages a day or some mad postingbot is flooding the Net with anti-Armenian hatred, then manually forging cancels for all those messages can become quite laborious.

Enter the cancelbot—a cancel program, "running in fully

automated mode," according to Depew, "with no human in the loop." A cancelbot automates the procedure of forging cancels for unwanted Usenet articles. Stationed at a Usenet news server, the cancelbot scans the incoming news feed for messages that contravene its programming. If set to search for Depew-authored messages, then any message from Depew in the news feed automatically triggers a cancel message.

There are few limits to the power of a cancelbot. One very early cancelbot program generated cancels for every single article emanating from a particular node. The tactic, dubbed the Usenet Death Penalty, is generally considered overkill of thermonuclear proportions and is reserved for those occasions when Usenet consensus deems that a particular news server administrator is out of control. A very rare event, indeed.

Most Usenet historians consider ARMM to be the first true cancelbot. All told, Depew wrote at least five versions of the program. The most notable were the first, ARMM-ANON Two Shot Turkey Shoot, and the fifth, ARMM5 (aka the ARMM cascade). On the first go-round, everything worked fine. On its trial run, ARMM-ANON canceled two anonymously posted articles appearing in the sci. hierarchy.

ARMM-ANON made history as the first cancelbot. But the release of ARMM5 created bigger ripples: the first cancelbot disaster. This new version of ARMM had a twist: it was designed to cancel an anonymous message and then repost it, along with an explanation alerting readers as to why the post had been deemed objectionable. Depew intended it as a public demonstration of cancelbot potential.

He succeeded more than he ever wanted to. Depew committed two of the most basic botrunner mistakes: he failed to test his program thoroughly before launching it, and then, after turning it on, he neglected to scrutinize its progress.

ARMM5 immediately headed straight into infinite recursion territory. A bug in its programming led it to treat its own re-postings of canceled messages as newly posted anonymous messages. Following its original orders, ARMM5 accordingly issued a new cancel message for the new post and then re-posted it, layering yet another explanation of its actions on top of the previous one. Within a few hours, more than 180 messages, each longer than its predecessor, poured into news. admin.policy. The stream of messages, today remembered as the ARMM cascade, presented as fine an example of Usenet spew as one could ever hope to see. It halted only after a news administrator phoned Depew.

Depew found himself in the unenviable position of a doctor whose cure happens to be worse than the disease. "My interest in canceling spew started at that moment," says Depew today. At the time, in a letter posted to news.admin.policy, he expressed contrition:

I offer my deepest apologies for this flood. I messed up badly. I made mistakes in both implementation and testing. That was truly a bone-headed implementation error! . . .

Julf's anonymous server seems to me to be contributing to the erosion of civility and responsibility that have been the hallmarks of the more traditional parts of Usenet. . . . I think it *is* important to demonstrate that Usenet *does* have a defense against a self-styled cyberpunk who refuses to cooperate with the rest of the net. Whether Usenet can find the *will* to oppose him remains an open question. I simply intended a brief demonstration of one defense mechanism.

I went into the lab to look for an anti-pathogen that would inhibit the growth of the pathogen. I found one—the Usenet Death Penalty. This was clearly dangerous stuff, so I tried to attenuate it—to improve its therapeutic index. The clinical trial was successful, at least in temporarily eradicating the pathogen from the patient's brain, but the patient unexpectedly suffered a severe allergic reaction, so I halted the test out of compassion.

DEPEW'S LEGACY

The ARMM cascade marked a major stage in the history of Usenet bots. It is, in fact, exhibit A in the prosecution's case against cancelbots. Depew still hasn't escaped the fallout— snide comments about ARMM in current Usenet debates are an enduring pox on his name. Programmers give no mercy to botched implementations.

But even though ARMM5's technical self-immolation of-fered ample evidence of the dangers of autonomous programs, that aspect of its performance was only a sideshow in the big-ger picture. The truly egregious factor for most of Depew's crit-ics was not ARMM5's foul-up, but Depew's decision to target anonymity. There was, and is, no consensus in cyberspace on the value of anonymity. By acting unilaterally, Depew had im-posed his own views on a community that prided itself on genuflecting to no ruler.

To many netizens and cyberpunks, the right to anonymity is a cornerstone of the libertarian approach to the Net. It is seen as one of the ultimate protective strategies for free speech in an authoritarian world. Impenetrable, cryptographically locked anonymity is touted as a defense against the snooping forces of Big Brother and corporate multinationalism, provid-ing the only safe cover from the Panopticon that sees all. By aiming his cancelbot at anonymity, Depew painted a bull's-eye on the principle of free speech and set himself up as an icon of the one thing that netizens hated most—censorship.

Censorship implied authority, and Usenet respected no

rulers. Who appointed Dick Depew guardian of Usenet? What gave him the right to decide what other people could or could not read? Years before the ARMM incident, a group of fundamentalist Christians had taken advantage of a loophole in an old Usenet newsreader program to do a mass cancel of "offensive" postings in the newsgroup alt.sex. Such religious busybodyness had been universally decried. What made Depew's actions any different?

To the Usenet hard-core, there *was* no difference—any act of censorship was by definition unacceptable. But Depew's critics were constrained by the same problem that hampered Depew's own attack on anonymity. Usenet had reams of etiquette, but no laws. Usenet had community norms, but no way to enforce them, other than by a concerted effort of news administrators cooperating together—a difficult thing to achieve.

ARMM5 stabbed at Usenet's sorest spot and, by extension, illuminated a core dilemma for the entire Net. How do you solve a widespread problem in a decentralized environment? What happens when community norms either do not exist or are inherently unenforceable? No matter whether the villain was Depew or Helsingius, the issue at hand was identical— Usenet had no effective mechanism for exercising authority, for resolving disputes, reining in malcontents, or ensuring that the greater good be served. Nor did any other region of the Net, excluding isolated MUDs. Transcending geographical boundaries, unregulated by government bureaus, constantly rebuilding itself and surging in unexpected directions, the Net obeys no master. But what price anarchy?

For the hard-core libertarians of Usenet, such freedom constituted a sacred trust. And if that was a problem for people, well, tough. If you didn't like anonymous posts, just delete them. If you found the discourse too harsh, go somewhere else. Individuals must take responsibility for their experience. Depew, on the other hand, had taken responsibility for everyone's experience. But of course, in a different way, so had Helsingius. Both were using software tools to extend their ideologies in the virtual world. Helsingius had his anon.penet.fi mailer daemons working overtime. Depew countered with cancelbots. The freedom inherent in the structure of the Net allowed them to exercise their bot surrogates with impunity.

ARMM's ignominious debut threatened a Darwinist dead end, a bot cul-de-sac. Instead, cancelbots thrived. Even after Depew had been reviled on both ideological and technical grounds, *more,* not fewer, cancelbots appeared.

The reason had to do with larger social issues than the concerns of Usenet's old guard. By the time of ARMM's arrival, the Net was no longer a children's sandbox. To borrow science fiction author William Gibson's term, 1993 was the year "when-it-all-changed" for cyberspace. Cyberpunk made the cover of *Time* magazine. The Web began its skyrocket ride to mass media glory. The words *email* and *Internet* became mainstays in the popular vocabulary.

All the fecund neighborhoods of the Net experienced dramatic population growth, and that growth led to new, hitherto unknown problems—like predators. As mainstream America woke up to the Net, Usenet's millions of readers suddenly became viewed as a captive market. Con artists, junk-mail advertisers, low-budget marketers, and political propagandists flooded Usenet like carpetbaggers. They saw an opportunity to get their message across at a low cost and with minimal effort. Why, a single spambot script could post identical messages in every one of Usenet's nineteen thousand newsgroups! How could one resist?

The old-school Usenet libertarians argued that unwanted speech is the price a free society pays for free speech. But even as they raised their free speech flags high, an ever-more-vocal majority of Usenet participants made it clear that they were willing to accept limitations on speech if that meant they could read their own chosen newsgroups without being required to endure intolerable levels of spam and pointless flamewars.

The concerns of a few First Amendment absolutists could not match the pressure exerted by greater social anxieties. As a choice between two evils, the contest was over before it started. Spam annoyed far more people than did cancelbots.

THE CABAL

Spam's defining moment came on April 12, 1994. Two Arizona-based lawyers, Laurence Canter and Martha Siegal, launched Usenet's first megaspam. In less than ninety minutes, they hit six thousand newsgroups with an advertisement offering assistance in the US Green Card Lottery—a chance for immigrants to the United States to qualify for a coveted work permit. Their assistance entailed little more than some help in filling out the application forms for participating in the lottery, a task for which Canter and Siegal planned to charge $95 a pop.

Spam, according to Usenet's Net Abuse frequently-asked-questions files, is technically pigeonholed as "the same article (or essentially the same article) posted an unacceptably high number of times to one or more newsgroups." Canter and Siegal were guilty of "excessive multiple posting"—a term later developed by Usenet's spam fighters to differentiate increasingly egregious forms of spam. (Precise definitions of *spam* were not formulated until after the green card escapade.) At the time, all people knew was what they saw through the lens of their newsreader software programs: the same crass rip-off in every single newsgroup they wanted to read.

Canter and Siegal weren't the first spammers to abuse Usenet, but as the Net Abuse FAQ succinctly put it, "they were so gothically clumsy about it, and so intent on making a buck, that people were terrified and infuriated."

Concerned Usenetters created the alt.current-events.net-abuse newsgroup—the predecessor to n.a.n-a.m—as part of a long-range strategy of monitoring and responding to spam attacks. In the shorter term, outraged Usenet residents expressed a more visceral reaction. They fought back.

They had weapons ready to hand: cancelbots. Depew had shown the way. Rahul Dhesi, a systems administrator in Texas, spent the first night after the initial green card spam hacking together a cancelbot script whose sole purpose was to expunge every trace of green card slime from Usenet. Even today, the lightning-fast response to the green card incident is remembered as one of the finest hours ever in the cancelbot defense of Usenet.

Suddenly, concerns over protecting free speech or exercising unilateral authority were shoved aside. To the independent saviors of Usenet, Canter and Siegal's blithe attempt to exploit Usenet for their own unsavory purposes presented a far greater threat than the prospect of undermining the First Amendment. Many Usenet regulars passionately believed that the Net offered a respite from the commercial pressures of the offline world—that somehow the wonders of the digital information era offered a future better than the past. Instead, they saw their edifice beginning to crumble, chipped away at the base by louts and savages intent on scrabbling for nickels and dimes.

Usenet won the first skirmish. Canter and Siegal were hounded off the Net. But there was no turning back the tide. As more people poured into Usenet, the flow of spam only increased. As one cancelbot operator noted, "The rate of spam is growing just as the Net is growing. And the rate of canceling follows suit."

The antispam resistance was anything but organized, but in the months and years after the green card incident, a gradual consensus on how to deal with spam began to emerge, centering on the activities of a loose group of cancelbot operators who shared tips and code with each other. Some operated out in the open, like programmer Chris Lewis. Others, like the Cancelmoose, cloaked themselves in near-total anonymity. All were volunteers who monitored spam and set into motion their cancelbot countermeasures according to an elaborate system of rules and regulations that were hammered out through open debate.

The evolving antispam guidelines they used to justify their actions expressed a distinct Usenet poetry, a mix of pseudo-officious jargon with its own crazy logic. Spam cancelers were urged to follow the $alz and cyberspam conventions, two sets of rules designed to help news administrators distinguish approved cancelbots from rogue vigilante bots. Then there was the Breidbart Index, a mathematical formula that took into account the number of newsgroups an article was posted to. As a hard-coded definition of *spam*, the Breidbart Index offered spam cancelers a handy defense when they were accused of subjectively censoring articles according to their own haphazard prejudices.

And accused they were. For even though the guidelines were included in numerous FAQs and were constantly being adjusted and tweaked in full public view, they were by no means unanimously accepted as legal writ by all Usenet participants. Some hard-line free speech activists angrily rejected every rationale put forth by the spam cancelers as self-serving bureaucratese. These dissidents accused the cancelbot operators of acting as a cabal that aimed to lord it over all of Usenet.

The members of the so-called Cabal scoffed at notions that they had any secret plan to rule Usenet, but they often acted in unison when facing a major spam threat. They also were careful to keep the actual code of their cancelbots to themselves. They didn't want malcontent users operating cancelbots according to spontaneous and uncontrollable whims. The programs were far too easy to abuse. In the text accompanying an approved cancel, such as the following, the cancelbot operators went to great efforts to stress their own aboveboard and legitimate reasons for taking action:

> From: Cancelmoose[tm]
> Subject: Spam Canceled
>
> This 190 message spam has been canceled from 179 newsgroups. It was not crossposted.
>
> Disclaimers: This is not being done because the message is an ad. This is not being done to censor a critic. This is not being done because I am offended by the words in the message. The poster has not been "silenced"—any and all non-spam messages from the same poster will remain untouched by me. . . . The commercial nature of this spam did not influence its cancellation. Spam is determined on number of newsgroups posted to, and not the content of the message.
>
> The $alz cancel. and Path: cyberspam conventions were followed.

The various conventions, indexes, and cancelbot guidelines developed in Usenet roughly paralleled the Robot Exclusion Protocol developed by Martijn Koster as a means of mitigating Web robot abuses. The Breidbart Index and $alz conventions were bottom-up responses to situations perceived to be community problems. And as much as certain parties decried them as unilateral impositions by a self-appointed techno-oligarchy, they still had the support of the only constituency whose voice really mattered on Usenet—the majority of Usenet news administrators.

"Usenet is a cooperative anarchy... [T]hat means it doesn't have a government, not that it doesn't have rules," says Dick Depew. "The rules are enforced in best anarchist tradition, by anyone who can convince enough news administrators that some particular rule can and should be enforced."

Enforcement, in a digital environment like Usenet, often entails no more than flipping a software switch. Usenet news administrators can decide whether or not their news servers will allow cancelbots to work by changing the software parameter that determines whether the server will respect cancel messages. Since most administrators believe that combating spam is more important than an absolutely purist free speech stance, "approved" cancelbots are free to work.

Another reason, of course, to keep the switch flipped that enables cancel messages is that individuals often have perfectly valid reasons to want to cancel their messages. Since Usenet software at the time mandated that anyone wanting to ban cancelbots would also end up banning all individual cancels, news administrators did not have much choice. And there was the rub. The system administrators were damned either way. Despite the best efforts of the cancelbot Cabal, cancelbot technology proved uncontainable. There was no way to ensure that cancelbots were always used wisely and appropriately, other than to change the system architecture so as to ban all cancelbots—a step that many argued could not be taken, given the constant battering on Usenet's doors by legions of spammers.

Just as was true in IRC and on the Web, Usenet provided yet another example of how agent/bot technology becomes only more accessible to the masses, not less.

Conventional cancelbots required twenty-four-hour access to the Usenet news feed, familiarity with Unix operating systems, and the ability to program. But to many observers, it was only a matter of time before some enterprising hacker would write a user-friendly cancelbot accessible to the millions of people with America Online accounts or Windows 95 personal computers connecting intermittently by modem to the Internet. Spammers already marketed Usenet-configured spambots. How long would it be before the Robert Returned of Usenet appeared, offering cheap Java cancelbots to one and all, for a mere pittance? After all, wasn't that one of the essential trends visible in the evolution of cyberspace—the creation and distribution of autonomous programs that amplified individual power?

Even if the cancelbot Cabal had managed to keep a lid on cancelbot technology, that still wouldn't have solved Usenet's problems. With their own powerful cancelbots at their disposal, the Cabal, ultimately, could do no more than soften the worst blows. They could not stop spam in advance, could not even slow down its malignant tumor-like swelling. Though the spam cancelers frantically strove to upgrade their techniques and to evolve their cancelbot scripts in step with spambots and other spammer tools, they could not keep up.

Hence the crisis that Usenet faced in the aftermath of the September 22 mass cancel attack. If indeed the intention of the mass cancel mastermind had been to demonstrate the hopelessness of defending Usenet with cancelbots, then the Machiavellian botrunner had succeeded.

But cancelbot vulnerability was only one of Usenet's problems. By the mid-nineties, Usenet, one of the oldest and biggest integrated forums for online interaction in the world, teetered on the verge of collapse. If it wasn't already broken, the day when it would shatter into a million splinters seemed imminent.

In 1996 a new visitor to Usenet might be hard-pressed to understand why, for so many early Internet explorers, Usenet had beckoned with all the promise and potential of a fresh new world. Usenet was awash in lost groups—newsgroups once ostensibly devoted to a particular topic that had long been abandoned by their original inhabitants. Only the hardiest of souls dared venture into a lost group to look for useful information or meaningful human contact. Amid solicitations for porn, rip-off scam come-ons, and endless content-free flamewars, here and there a lonely cry for help might bubble up, unanswered, unheard.

INVASION OF THE ROBO-MODERATORS

Neo-Tech: A noun or an adjective meaning *fully integrated honesty* based on facts of reality. Neo-Tech creates a collection of *new techniques* or *new technology* that lets one know exactly what is happening and what to do for gaining honest advantages in all situations. Neo-Tech provides the integrations in every situation to collapse the illusions, hoaxes, and all other forms of mysticism manipulated by the parasitical elite. . . .

If Neo-Tech type business was the order of the day, and parasitic value leachers were socially ostracized, not only would stagnating unemployment and non-occupation not exist, but competitive opportunity would ensure the availability of life preserving biological immortality within several years.

—"APO Takeover Manifesto"

In the spring of 1996, a cursory glance through the titles of the topics being discussed in the newsgroup alt.philosophy. objectivism baffled the casual visitor. Neocheaters? ZonPower? Takeover manifestoes? The pros and cons of robo-moderation? None of it made any sense, and hardly any of it seemed related to the topic at hand: the libertarian strand of philosophy known as objectivism.

As set forth in the writings of Ayn Rand, author of the novels *Atlas Shrugged* and *The Fountainhead,* the core principles of objectivism are reason, rational self-interest, and a laissez-faire free market. On the surface, then, you might logically assume that a newsgroup devoted to such ideas would be a forum for reasoned, careful debate. It's no surprise when the witches and warlocks in soc.religion.paganism let their oratory run wild or when the passions of the stuffed furry animal enthusiasts in alt.sex.plushie bubble over. But a congregation of Ayn Rand fans consumed with bizarre sci-fi psychobabble?

On Usenet, anything is possible. Even objectivists can act irrationally. That spring, alt.philosophy.objectivism, a newsgroup that should have been a case study in online rules of order, instead found itself seething in chaos and confusion. The subtleties of Ayn Rand's exaltation of the principle of "individual rights" had no chance in the wild battle for newsgroup dominance being waged between an old guard of classic objectivist thinkers and a strange new objectivist splinter group— the Neo-Techs.

Dismissed as a really bad ongoing joke or condemned as a sinister Scientology-like cult, by March 1996 the Neo-Techs had been plaguing the a.p.o newsgroup for more than a year with profuse postings praising the merits of *ZonPower*—a hazily defined self-help ideology. According to the Neo-Techs, if properly embraced, ZonPower would deliver "open-ended life, prosperity, and happiness to conscious beings throughout the universe." But if anyone attempted to debate the Neo-Tech point of view, the Neo-Techs responded by gangposting en masse, swamping the newsgroup with hundreds of pseudoscientific, jargon-laden, repetitive propaganda posts.

In the best of times, a Usenet newsgroup is a disorienting place. Hundreds of conversations take place simultaneously. Topic threads can continue for months, often ranging far afield from the original comment or question that initially set off debate. The Neo-Techs magnified normal Usenet disarray by at least an order of magnitude. They crashed the a.p.o party with hundreds of "templates"—prepackaged rants that they reposted thousands of times to the newsgroup, in any thread they deemed appropriate. They even forwarded posts from their own private Neo-Tech mailing list discussion group directly into the a.p.o newsgroup, creating a daily deluge of messages that had little relevance to any ongoing conversations, aside from those that focused on combating the Neo-Tech menace.

On March 16 the Neo-Techs delivered their ultimatum—they released the Neo-Tech APO Takeover Manifesto to alt.philosophy.objectivism. A four-hundred-line screed posted five or six times a day to the newsgroup, the manifesto made clear the Neo-Tech intention to rid the newsgroup of all unbelievers, or "neocheaters," to use a favorite Neo-Tech insult. The ensuing flamewar threatened to push an already tottering newsgroup over the brink.

Though the total number of individual Neo-Tech posters was small, mass numbers weren't required to ruin the a.p.o newsgroup as a forum for meaningful conversation. As Betsy Speicher, a prominent figure in the objectivist community, noted, "one person—or a half-dozen persons—can do a lot of damage in an unmoderated environment." And they did. The Neo-Tech input threatened to turn alt.philosophy.objectivism into a classic lost group.

To the a.p.o regulars, the Neo-Techs disgraced Rand's legacy: they were "spammers and loons" intent on making a mockery of objectivist ideals. Even worse, after a little digging into the Neo-Tech background, it seemed obvious to the regulars that the Neo-Tech hullabaloo was intended mainly to publicize the products of the Neo-Tech Publishing Company—an expensive line of Neo-Tech/ZonPower self-help books boasting grandiose titles like *Riches from Another World*.

The Neo-Techs disregarded all criticism. They believed they had every right to do as they pleased. After all, wasn't free speech one of the sacred tenets of objectivist philosophy?

That indeed was the question. As the a.p.o regulars contemplated their options, they found themselves confronting a powerful contradiction. Usenet's own celebration of free speech had ended up rendering huge swaths of the network effectively useless. Hundreds of newsgroup communities faced the same problem, but none, perhaps, demonstrated the paradox as dramatically as did the objectivist newsgroup, where free speech advocates sought desperately for an effective means of stifling Neo-Tech expression. When the objectivists sought succor from a bot, they sent a signal to all other newsgroups that philosophical objections to bot involvement in newsgroups simply didn't hold up.

The objectivist regulars established a new newsgroup—humanities.philosophy.objectivism. In contrast to the old newsgroup, h.p.o would be moderated. It would have rules delineating acceptable posting. And the rules would be enforced by a new kind of Usenet bot: the modbot.

This was dangerous ground for libertarians, many of whom contended that the term *moderation* was just another excuse for imposing censorship. But Betsy Speicher, one of the leaders of the modbot movement, shrugged off such criticism. Since the rules were part of a preset policy, she argued, anyone who voluntarily participated in the newsgroup de facto consented to the regulations. The new newsgroup would not replace the old newsgroup—and, to be sure, alt.philosophy.objectivism still exists today, as unreadable as ever. Ever zealous to keep their own hands clean, the objectivists even turned to two outside figures—Dick Depew, of ARMM fame, and Tim Skirvin, the author of the Usenet Cancel FAQ—for help in supervising modbot operation.

The objectivists weren't the only group to see bot moderation both as an escape from "net-abusers" and as a way of avoiding the sticky questions of subjective censorship. The newsgroup soc.religion.quaker also experimented with modbot moderation, believing that using a bot to enforce posting guidelines allowed them to conform to the Quaker religious precept that forbade "putting one man above another." Let others argue that behind every bot must stand a human; such philosophical quibbling paled before the threat of ZonPower and the like.

As opposed to a cancelbot, which eliminates Usenet articles after they have already been posted, a modbot reviews articles before they ever see the light of Usenet day. The h.p.o modbot, as envisioned by Speicher and her colleagues, rejected articles that contravened certain posting criteria. Spammers would be banned. Crossposting—the practice of simultaneously posting a single copy of an article to multiple newsgroups, so all responses to that article in any newsgroup appeared in all the others—would not be allowed. And most controversially, any post that contained Neo-Tech buzzwords such as *neocheater* or *fully integrated honesty* or that in any way promoted Neo-Tech wares would be instantly rejected. Though censorship based on content constituted original sin in the eyes of many Usenet hard-liners, the objectivists ignored their qualms. Today the objectivist refuge exists happily in its modbot-guaranteed safety, free of Neo-Tech blather.

The Neo-Tech vignette works well as a mini morality play about the clash between free speech ideals and anarchic reality and about the stresses caused when scam-artist greed capitalizes on the popular desire for human communication. The modbot solution is also one more example of netizens seeking technofixes for social problems.

The objectivists were hardly alone in seeking modbot help, but the h.p.o modbot was hardly a perfect solution. In fact, it had one giant gaping loophole. The modbot proved helpless against faked messages, or "spoofs." Dedicated opponents of the newsgroup could still take advantage of Usenet's structural flaws to cause as much damage as they desired. Spammers could easily forge new identities, allowing them to sneak past modbot sentries.

For some newsgroups, particularly those that edged near discussion of volatile political and cultural issues, forged postings posed as big a threat to discourse as any Neo-Tech invasion. Perhaps the best and most notorious example of what happened when forgeries ran rife was visible in soc. culture.russian, a newsgroup established to provide a place for the discussion of Russian affairs. For years, a set of shadowy characters who could have leaped straight from a Dostoyevsky novel plagued soc.culture.russian with an endless flurry of forged postings. Vile character assassination, muddy intrigue, conspiracy, and violently feuding factions—the atmosphere of s.c.r eerily reflected the real-world political and cultural chaos of Russia itself. No ordinary, objectivist-style modbot would ever have been able to cope.

Different circumstances, then, called for a different kind of modbot. Luckily, Usenet's distributed decision-making structure encouraged multiple solutions to multiple problems. Igor Chudov, a United States–based Russian émigré, hatched his own robo-moderation plan, one that incorporated cryptographic authentication as part of its defenses.

Chudov's bot integrated cryptographic security measures on two levels. First, would-be posters were encouraged to sign their posts with cryptographically locked digital signatures that incontrovertibly affirmed their identity to anyone who had the proper tools to read the signature. Chudov's robo-moderator had those tools, along with an automatically updated list of signatures that it could check each new article against. Second, if a strange newcomer posted an article to the group, the robo-moderator forwarded the message for approval to the group's human moderators, signing the message with its own digital robo-signature to ensure that nobody infiltrated the moderation process at that level.

At first glance, the robo-moderation scheme seemed absurdly complex. But it worked. The newsgroup soc.culture.russian.moderated is now a peaceful place, and the intricacies of the scheme are well removed from the regular user. Onlookers beset with similar problems studied the Chudov system. By late 1996, soc.religion.paganism—also the target of regular harassment—had voted to install robo-moderation, and other newsgroups were expressing interest.

Meanwhile, the h.p.o modbot solution spread to other newsgroups, with modifications implemented to take account of local circumstances. And still other newsgroup communities authorized their own homegrown modbots. Increasingly, for any newsgroup beset by some perceived problem, robo-moderation seemed like the right answer. A host of religious newsgroups, motivated in large part by massively destructive crossposted flamewars initiated by a single person, adopted modbot systems. So did technically oriented newsgroups, regional newsgroups, and many others. The group sci.geology got a modbot. So did misc.education.home-school.christian.

Modbots proved that the solution to Usenet chaos did not require centrally operated cancelbots ranging the length and breadth of all of Usenet. Purely utilitarian considerations determined whether Usenet's equivalent to IRC's channel protection bots—the robo-moderators and automoderators and modbots—succeeded or failed. Did it work? Was it hard to install? Would the upgrade be adopted by enough people to give it critical mass?

"Any node can submit an article, which will in due course propagate to all nodes." So spoke Tom Truscott in 1980, describing the fundamental principle that underlay the Usenet architecture. Modbots, too, propagated from node to node, according to their fitness with various environmental needs. With no central planning and no unified direction, the modbot solution flourished. In the absence of government, a thousand tiny governments sprang up, each with its own bot-enforced constitution.

FERTILE GROUND

Even as modbots spread across Usenet, breeding like rabbits, some of the old cancelbot operators, recognizing that the cancelbot approach offered no long-term security, focused their energies on a new target. The Cancelmoose, in particular, set its sights on a systemwide solution that would negate both the problem of forged cancels and the criticism of cancelbots and modbots as tools of censorship. While others sought local solutions, the Cancelmoose went global.

It envisioned a future in which the individual Usenet reader had total control, a future in which the reader—not some self-appointed Cabal or distant news administrator—decided what constituted spam and what was acceptable speech.

The Cancelmoose was convinced that the key to reader empowerment lay in the software program the human reader used to sift through Usenet's millions of articles. What if the reader could painlessly configure that software to accept cancel messages only from the original author of the message to be canceled or from someone the reader trusted? With the proper incorporation of cryptographic authentication, one could dismiss spoofing or forgery worries.

In late 1995 the Cancelmoose retired from active cancelbot operation and hacked together a solution, called NoCeM (pronounced "no-see-um"—as in, no-see-those-bad-messages). NoCeM is a newsreader accessory. Once installed, it filters the news feed at the point where the individual reader interacts with Usenet, according to whatever parameters the reader has set. If lazy, the reader can simply let the default recommendations operate—for NoCeM, that means any cancel messages issued by the Cancelmoose's AutoMoose spam detector or by Chris Lewis, Dick Depew, and the rest of the long-established cancelbot operators.

But the reader doesn't *have* to accept the AutoMoose as Usenet gospel. Anyone who believes that Lewis et al. are traitors to free speech can turn off all the defaults and receive every message that anyone posts. Or the reader can turn to other authorities for filtering advice. The biases of any political, economic, or cultural institution can be incorporated at will into NoCeM. Tired of blasphemy on the Net? The likes of Pat Robertson or the Christian Family Association can create a bot to patrol the Usenet news feed for unholy utterances. A NoCeM newsreader will allow readers to configure it to respect the Christian bot's revelations. Sick of libertarian propaganda? Set your NoCeM newsreader to accept cancel messages from the Ayn Rand watchbot. This isn't censorship at all—just free individuals acting according to their own choices.

By the end of 1995 the Cancelmoose had a prototype of its solution in working order. But at first, only a few people implemented it. NoCeM was a geek-only technofix, technically complex and applicable only to computers running Unix operating systems. It simply wasn't an option for the Macintosh or Windows masses.

Then came the great mass cancel assault of September 1996. The elimination of thousands of legitimate postings exposed the Achilles' heel of the cancelbot solution once and for all. But in the aftermath of the disaster, as concerned Usenet citizens pondered Chris Lewis's challenge—do we cave in, or do we stand firm?—a funny thing happened. The question of who had masterminded the mass cancel attack received short shrift. It was far more important that Usenet move forward. Suddenly, more people began paying concentrated attention to the NoCeM newsreader than ever before. The remnants of gift economy idealism kicked back into gear. How could NoCeM be improved? Made more accessible? Adapted to other computing platforms?

It is still not clear whether the NoCeM solution will succeed. Commercial newsreader software companies may come up with their own proprietary schemes that swamp the altruistic offerings of the Net. Or some other hacker may propose a better solution. But NoCeM's success or failure is irrelevant. Something will fill the void.

Usenet, even as it faced its worst crisis, even as it was close to shambles, even as it was collapsing, was being rebuilt with an industrious self-modifying energy. On the grassroots level, in a thousand different newsgroups, a thousand different modbots blossomed. And at the same time, dedicated programmers were tackling the broader issues of remodeling the overall

Usenet infrastructure itself, revamping the newsreaders and the news transport protocols. From top to bottom, the network was in flux, adapting and changing.

Usenet displayed an almost Nietzschean vigor. That which it could endure only made it stronger. The worse the crisis, the more energy was devoted to finding solutions, and in a decentralized system that promoted multiple experimentation, workable answers enjoyed the freedom to prosper. Anarchy didn't just promote chaos; it provided fertile ground for new approaches.

Usenet isn't broken—it's in ferment.

THE TECHNODIALECTIC

> At each period of growth all the growing twigs have tried to branch out on all sides, and to overtop and kill the surrounding twigs and branches, in the same manner as species and groups of species have at all times overmastered other species.
>
> —*The Origin of Species,* CHARLES DARWIN

In virtual West Texas, Buford and Nurlene Moot no longer push pencils in their Point MOOt offices, dispensing government contracts and welfare largesse. The bum and hooker bots are silent. The streets they once roamed lie empty. Even Cthulu's reign of terror is over. No more does the great slime-covered monster strike horror into the hearts of Point MOOt residents. And the hordes of Barneys? They are nowhere to be seen. Their song has been stilled.

Few will lament the demise of the Barney bots. Their excesses threatened to rip Point MOOt apart. But the Barney plague did not bring down Point MOOt. To the contrary, it was the very success of Allan Alford's ambitious reality modeling experiment that proved to be the city's doom. Popularity is dangerous on the Net, where gold rush fever can strike in seconds and massive crowds will stampede at a moment's notice. As the Internet exploded over the summer of 1994, Point MOOt caught part of the blast. After a few admiring articles in the press and a vigorous word-of-mouth campaign in newsgroups and mailing lists, huge waves of immigration overwhelmed the Point MOOt infrastructure.

The ACTLab workstation that housed Point MOOt failed to cope with the frenzied hustle and bustle. As the summer wore on, the MOO slowed to a sluggish crawl, beset with nearhourly system crashes. Alford begged his faculty advisers in vain for an infusion of new funds. Finally the archwizard had to face a reality that could not be modified by a software tweak. Barely six months after it officially opened to the public, Point MOOt closed its telnet doors.

The source code is still available, a set of files stored on a hard drive in Austin, Texas. Today the bots of Point MOOt slumber in suspended animation, cryogenically awaiting some change in Alford's fortunes. As computer memory and processing power become cheaper, Alford hopes that one day it may be feasible to operate Point MOOt from his home computer, that one day the bots of Point MOOt will live again.

We haven't heard the last from Barney. But even if the silly dinosaurs never infest cyberspace again, even if Point MOOt is no more, for now, we would do well to remember the legacy of Point MOOt. Bots never do exactly what we intend them to do. Even worse, behind every helpful Buford lurks an evil Cthulu, net sack in hand.

Point MOOt, albeit short-lived and self-contained, nonetheless waved a warning flag. A plague of Barneys is no accident—it is an inevitability. Similar catastrophes can occur in any online neighborhood. The Net is poised for an invasion of purple dinosaurs singing at the top of their lungs, threatening to drown out all of cyber civilization.

"That's right," says Alford. "The Barneys are coming. And you might not have your Barney Blaster."

< >

But are guns the answer? Any experienced observer of online dynamics knows that simply arming the citizenry doesn't always make the most sense for coping with foxbots in the cyber chicken coop. Reaching for a more powerful weapon invites escalation. In IRC the channels with the strongest defenses attracted the most concentrated attacks. In Usenet the employment of cancelbots in the service of newsgroup peace and prosperity led directly to their use as tools for chaos. A suit of armor and a battle mace are no ultimate panacea.

Indeed, if one is looking for panaceas in cyberspace, then one will search for a very long time. There is no ultimate security in an open-ended, decentralized, and distributed network. The ground is always shifting. But that is not a flaw in the system that needs to be fixed. It is a feature. The impossibility of absolute safety is a reflection of the greatest strengths of networked environments: flexibility, innovation, and *flow*.

Usenet offers the clearest demonstration of the principle. Decentralization in Usenet fostered anarchy—a mad jumble of forged cancels, flamewars, and lost groups that drove thousands of Usenet participants to the edge of despair. But Usenet also fostered experimentation, encouraged trial and error, and catalyzed endless new approaches. Since, in practice, every Usenet newsgroup constituted its own independent community, each group could test out its own solutions, its own rules, its own bots.

The result has been the living, breathing incarnation of an idea most eloquently expressed in another time and place. "Let a hundred flowers bloom, let a hundred schools of thought contend," the leaders of China told their people in the 1950s. Strength comes from diversity, from multiplicity. Out of chaos, possibility. Usenet's zoo of modbots, its smart newsreaders and system architecture overhauls, its thousands of newsgroups each pursuing their own path to nirvana—taken all together, they are the signs of a healthy, contending, blooming system.

Usenet is just the most obvious demonstration of the hundred flowers principle. The Net, both in its entirety and in each of its microcosmic composite sectors, exhibits the same fundamental life spirit. The fractured, asynchronous, yet parallel processing power of a distributed, decentralized environment provides answers to every dilemma.

"Things like [the Net] tend to be self-balancing," says David Chess, a computer security expert at IBM. "If some behavior gets so out of control that it really impacts the community, the community responds with whatever it takes to get back to an acceptable equilibrium. Organic systems are like that."

On one level, to think of the Net as an organic system is to push at the limits of metaphorical sense. Computers are the antithesis of organisms. Nature is not digital. When networks grow, it is not of their own accord, as nurtured by warm sun and soothing rain. It is because some human, somewhere, flipped some switches, plugged in some cords, and hacked some code. When a network like Usenet or IRC or the Web runs into problems or exceeds human comprehension, the first impulse of the indigenous technogeeks in each environment is to concoct some form of *artificial* technofix. What is organic about a software patch, or a bot?

Well, of course, humans are organic, and one can argue that anything we do or create is therefore part of an organic process. Human plus bot equals organic cyborg. Humans plus bots plus Net equals an unimaginably complex, multicentered collective of interlinked cyborgs—an organism greater than the sum of its parts.

Furthermore, humans seem unable to interact with any aspect of their environment without recklessly wielding metaphors, without wantonly anthropomorphizing and personifying. Whether or not the Net is truly organic, whether or not bots are truly autonomous new creatures, we will continue to think of them in such terms and accordingly shape and guide their development.

If the Net is defined as an organic combination of hardware, software, and humans endlessly self-balancing, then that very act of seeking balance is the life force that guides the evolution of the Net's new species: its bots and agents, its spiders and eggdrops and auto-meese. The environmental fitness of these new species is determined by the endlessly blooming and contending Net, always in search of solutions that will work within a technologically delimited reality.

Call it the technodialectic—a herky-jerky process of programming propagation, an emergent property of the Net, cyberspace's answer to Darwin's theory of natural selection. The technodialectic resolves the problems caused by bots and humans, and as it does so, it propels bot evolution onward. It's a peculiar kind of evolution: unnatural selection, a survival of the fittest program, determined not by nature but by the interaction between human and computer.

It is dialectical because every technofix generates a new problem, which in turn requires a new solution, which in turn is undermined again. For every search robot, there is a spambot; for every spambot, a defensive mailbot; and for every mailbot, a newer, smarter spambot. And so on.

Not all survive in the bot battle for life. The technodialectic implies an unending process of shakeout. Certain bots prove better than others and then propagate, from node to node. Good bots, useful bots, proliferate. Useless bots—too complicated, too ineffective, too easily thwarted—become extinct.

Usenet offers the most obvious manifestation of the technodialectic in action, as it applies to bots. The spread of a successful modbot from newsgroup to newsgroup, morphing according to the exigencies of the local terrain as it travels, is a classic case of technodialectical flow. But Usenet is not unique. The technodialectic courses through every Net ecology.

< >

On IRC, annoybots impelled the creation of channel protection bots. Channel protection bots inspired clonebots, collidebots, and floodbots. Each new iteration of warbot required a corresponding defensive improvement. Then the IRC server administrators banned all bots.

"Of course, this led to the classic battle of coder versus operator," says IRC hacker Chris Piepenbring (aka Hendrix). "The coders change their bots to look less like bots and more like regular IRC users. Then the operators find new ways to detect them. Then the coders patch up their bots to avoid them, et cetera. It's like a survival of the most clever."

At first glance, the ban on bots flew in the face of the technodialectical imperative. An anti-bot blacklist, enforced by canceling botrunner login accounts, is a raw application of power. For the technodialectic to work best, propagation must be voluntary. Solutions are not imposed from the top but evolve from the bottom to fit particular needs.

Of course, no ban is absolute; clever bots will always find a loophole. But even as IRC administrators during the mid-nineties ruthlessly hunted down bots and botrunners, other contingents of the IRC community sought out a different, less authoritarian solution to the bot problem. IRC's loose structure facilitated the creation of new IRC networks—just as anyone could start a channel on the original IRC network, EFNet, so too could any group of people start a new IRC network of their own.

And so they did. Motivated by disgust at the constant political bickering of EFNet, fleeing the chaos caused by the botwars, and convinced that they could do better if only they could start with a clean slate, successive bands of IRCers created new, smaller IRC networks with names like Undernet, DalNet, and NewNet. In the process of setting up these new networks, the latter-day IRC pioneers incorporated modifications in the system software that encouraged different user behavior and different varieties of bots.

The largest of these new networks, Undernet, eliminated the messiness of endless fighting over channel ops by establishing a single centrally run channel registration bot—the Xbot. The Xbot permanently registered all channels in the name of their original creators, thus obviating the need for an individual bot to watch over each separate channel.

The creators of the Undernet also instituted a technical procedure called time stamping. Time stamping made nick collisions impossible. In a single stroke of protocol upgrade, collidebots and clonebots became extinct. Time stamping soon became adopted by other networks, including the original EFNet, spreading in the same node-to-node style as a successful Usenet modbot. So even as the original network spawned new subnetworks, innovations in the subnetworks traveled back upstream and were incorporated in their forebears. The technodialectic flows in every direction at once.

But just as the Usenet hard-core libertarians attacked modbots as tools of censorship, so did the IRC old guard immediately criticize the centrally operated Xbot as an unwelcome central choke point. Instead of just being able to create a channel whenever you wanted, now you had to register it with the Xbot. Whoever controlled the Xbot ruled the Undernet.

Such criticisms were moot. If you didn't like the Undernet, then you could stay on EFNet or sample the wares of NewNet or DalNet. Multiple solutions, multiple approaches, multiple choices.

A more compelling criticism was that the proliferation of new nets fragmented the IRC community. This was not a problem for the more popular channels—in fact, it could even be a boon, reducing congestion and making conversations more manageable. But it did make life harder for the smaller, niche channels. In a world of endlessly subdividing networks, achieving the critical mass of participants necessary to make a chat room take off could be challenging. What happens when all the people interested in the cartoon show *Animaniacs* are divided up among four or five different nets? Isn't that a step backward?

Perhaps. But it's also a step forward. The constant bifurcation of networks created a need for a new kind of bot—the channel relay bot, or linkbot. A linkbot connects channels on different IRC networks together, relaying conversations back and forth so that inhabitants of both networks get the benefit of a larger community. Of course, channel relay bots then themselves become the targets for various forms of warbots and have to incorporate defensive measures. And so it goes.

The bigger the network, the more widely distributed, the more open-ended, the better the chance for the technodialectic to flourish. Even though one can see rudimentary traces of the technodialectic in a MUD like Point MOOt—Barney bots were followed by Barney hunters wielding Barney Blasters, which in turn led to plagues of Barneys and the arrival of Barney hunters armed with >@nuke< commands—the isolation of a MUD insulated it from the larger flow. The technodialectic could not save Point MOOt.

But it will save the Web, or at least sustain it, in the face of the oncoming wave of Web botwars. As judged by the criteria of open-endedness, nothing compares to the Web. No better arena for the technodialectic could be imagined. The Web is the ultimate network, absorbing everything it comes into contact with. IRC chat rooms and Usenet newsgroups can be accessed directly from a Web browser and are often linked explicitly to Web sites, creating a seamless fabric of online communication and contact. Even MUDs are evolving into Web-compatible forms. And it doesn't stop there. Commerce, culture, politics—the Web will connect everything.

The rhythms of the technodialectic are discernible everywhere on the Web. The growth of the Web led to Web robots, which led to search engines, which inspired spamdexing (as well as custom filters to block the advertisements that sprouted at search engine sites). Parasite metacrawler bots soon piggybacked on the efforts of the original Web robots. Web robots and search engine indexing programs are constantly being upgraded to repel the latest spamdexing tricks. And the spamdexers keep getting cleverer.

Meanwhile, webmasters and robot authors are engaged in a constant leapfrog battle of HTML point-counterpoint. The Robot Exclusion Protocol has proven inadequate to the pressures exerted by irresponsible bot owners, and in reaction, webmasters search for ways to defend against rogue bots. For example, they may require that every robot or browser requesting a document or file from a Web server identify itself first with a special code. No identification, no entry. Robots that misbehave are automatically banned from future visits. In such a fashion, recommended etiquette becomes infrastructural architecture.

On the Web, as elsewhere on the Net, for every negative bot action, there will be a positive bot reaction. And likewise.

But is that progress?

< >

There will be no ultimate comprehending of the Web—one main reason, of course, why we humans need bots in the first place—to help carve out some slice of sense for our puny brains. Already too much for any human mind to master, the Web's future complexity will boggle description. That fact can be unsettling. The Web's vast chaotic morass will be a sheltering friend to spybots and spambots and warbots of all descriptions. It will infinitely frustrate any single engineer who believes that a perfect solution can be constructed.

Engineers and programmers have a tendency to believe in the possibility of near-perfect solutions. Programmers who have reduced the world to ones and zeros are often unwilling to accept ambiguity and incompleteness. The structure of the computer aids and abets this mind-set. The advance of computer technology makes a strong case for "progress." But computers also breed hubris. A computer will do what you tell it to do—it executes *instructions*. Slavishly at the beck and call of the programmer, the computer fuels technogeek confidence, invites delusions of grandeur, dazzles with the promise of ultimate archwizardry. If the computer, or the network created out of linked computers, isn't responding as one wishes, then the instructions must be incorrect.

One computer will do as it is told. But a million computers linked together, responding to the needs and desires and obsessions of a million people, will not. A network is inherently unstable. The Web will be ever restless. It is at no one's beck or call. And no matter how many times software protocols are patched or how many loopholes are plugged, no engineer, or team of engineers, will ever finish the job. The Net may be self-balancing, but it is never in balance. No organic system is truly in balance. It may strive for balance, it may continually seek equilibrium, but it will never actually achieve it.

Indeed, constant conflict is as essential to an organic system as is the need for balance. Charles Darwin chose to describe the evolutionary journey as a "battle," for good reason. Conflict, for Darwin, was essential to evolution. The struggle for life is a bloody business. Species overmaster other species. Survival is the issue at hand—not progress toward some perfect form, some ultimate goal.

Darwin was frank. In *The Origin of Species* he declared that "no innate tendency to progressive development exists." Evolution just happens—it isn't headed anywhere, except onward. The theory of natural selection explains only how organisms adapt through time to changes in local environments. It does not mandate that those organisms become smarter or stronger or faster or lovelier. It just demands that they survive.

On a good day, the flow of the technodialectic on the Net seems to imply progress. Smart newsreaders, modbots, channel protection bots, linkbots, and mailbots—Breidbart Indexes and Robot Exclusion Protocols, Undernets and NewNets—are encouraging examples of what David Chess calls "systems designed to do the right thing in the presence of the prevailing level of unethical activity." Systems designed to prevent bot misdeeds. On a good day, the technodialectic will ensure that the "right thing" happens.

But not always, and not forever. The technodialectic may mute chaos, but it will never resolve it. For the struggle for bot life is also about survival, and not about a search for stability. To believe in safety is to fool oneself. There is no resolution—

just unending process. And there is no ultimate hiding place, no secure refuge, for anyone—bot, human, or cyborg—in the midst of that process.

< >

The technodialectic is no security blanket. It is only a thin strand of gauze fluttering at the edge of chaos, bending and weaving with every gust of disorder. And its ultimate direction always leads further into chaos, provoking entropy even as it ensures local order. No matter how deft our helpers are, no matter how cunning our daemons become, we cannot solve that problem once and for all. There will always be more chaos along the way. The bad bots will just keep on coming.

"Bots are becoming more and more complicated," says IRC's Int3nSiTy. "If you went away for a few months and came back onto IRC you would probably be lost. Who knows what we will be able to do in a few more months?"

Who knows indeed?

When Alex Cohen, the onetime chief technical officer at the McKinley Corporation, was in ninth grade, he read "Microcosmic Gods," a Theodore Sturgeon science fiction short story that has kept him thinking ever since.

"Microcosmic Gods" is the tale of an engineer so bored with engineering that he creates a race of tiny intelligent beings to do all his work for him, to solve any problem that he poses. He soon becomes rich and powerful by profiting from their inventions and is considered a threat to the world's military and financial powers. The real threats, of course, are the hardworking creatures—the Neoterics. The engineer controls them by ensuring that each individual's life span is fleetingly short, but even so, their strength far outshadows his.

"The question we have to ask ourselves," says Cohen, "is, are we becoming microcosmic gods?"

Are we using our bot helpers to extend our power in the virtual realm to the point that our ability to inflict our will becomes godlike? And how do we restrain a world full of gods? And what happens when our helpers finally throw off their chains and sever their cyborg links? Current robot incarnations may not seem as threatening as Sturgeon's little Neoterics. But the bot climb to power and glory has only just begun.

Science marches on. Before the McKinley Corporation lost out in the search engine wars and Cohen was forced to move on to greener pastures, he had already started daydreaming about incorporating real processes of natural selection into the next version of his Wobot Web robot. By using the promising AI technique of genetic algorithms, he would build the capability for constant mutation into the Web crawler.

"It will live or die, depending on how good a job it does," says Cohen. "The ones that do the best job live, and the ones that don't die. Its food will be information. If it gives me good information and categorizes things accurately, it lives."

The prospect of a mutating Wobot raises grand hopes of artificial life for our menagerie of daemons and bots and agents. But to endow such programs with true reproductive autonomy would not only cut the cyborg bond between humans and bots but also greatly expand the possibilities for bot-caused anarchy by orders of magnitude. Once annoybots and Websnarfs and modbots really start to run wild, current levels of disequilibrium will seem like mere ripples on a pond on a windless day.

Cohen is convinced that the emergence of artificially alive bots on the Net is inevitable. "Think of it this way," he says. "The Net is an environment. There is not a single environment on the earth that hasn't been invaded by life. It's only a matter of time before this new environment gets invaded."

The word *invasion* has a negative connotation, but Cohen isn't alarmed. The prospect of bot-induced destabilization is nothing to be afraid of, he contends.

"Ideally, the Net shouldn't be stable," says Cohen. "It should surge back and forth. For it to be a good Net, it should be prone to incompleteness and breakdown."

Cohen pauses, and smiles.

"Otherwise it is not a good place to be."

GLOSSARY

algorithm A rule of procedure for solving a mathematical problem. In computer science, an algorithm is a sequence of steps, or instructions, aimed at accomplishing a particular result.

annoybot On IRC, an autonomous program intentionally designed to disrupt chat rooms.

anonymous remailer A Net-connected computer that strips identifying information from an email message or Usenet post and then forwards it on to its intended destination.

applet A small program written in the Java programming language. Applets are designed to add interactive functionality to the World Wide Web.

automoderator See *modbot*.

AutoMoose A cancelbot, written by the Cancelmoose, that is able to detect spam and to issue messages to cancel it.

avatar In an online context, a graphical representation of a human, meant to interact with other such human representations in a three-dimensional domain.

botrunner   A person who operates a bot.

Breidbart Index   A mathematical formula for determining spam. The Net Abuse FAQ defines the index as "the sum of the square roots of how many newsgroups each article was posted to." According to the FAQ, "If that number approaches 20, then the posts will probably be cancelled by somebody."

cancelbot   On Usenet, a program that issues cancel messages for Usenet articles.

channel   A chat room on IRC.

chatterbot   A bot capable of carrying on a conversation with a human.

client/server   An adjective describing the network system by which individual "client" computers request and receive files, data access, or processing power from a more powerful "server" computer.

clonebot   On IRC, a bot capable of duplicating itself multiple times.

collidebot   A clonebot designed to kick humans or bots out of an IRC channel through the technique of nick collision.

EFNet   The largest and oldest IRC network, consisting of a collection of IRC server computers.

Eggdrop bot   Robey Pointer's channel protection bot for IRC that is both a helpbot and a defensive bot.

expert system   An artificially intelligent computer program that attempts to duplicate the expertise of a human in a particular subject area.

fuzzy logic   An AI technique designed to handle concepts of partial truth—that is, values that are neither absolutely true nor absolutely false. Fuzzy logic was originally conceived for use in natural language processing.

genetic algorithm An algorithm that is designed to mutate, breed, and spawn new, improved algorithms based on its success in solving a particular problem.

header In an email message or Usenet article, a sequence of data containing identifying information such as sender's name and address and the transport route of the message.

host A server computer from which Web pages, data, and files can be retrieved by Internet users.

HTML Hypertext markup language: the formatting language of Web documents.

HTTP Hypertext transfer protocol: the original communications protocol of the Web. HTTP enables browsers to connect to Web servers, and servers to communicate with each other.

infinite recursion A mathematical term for a procedure that repeats itself an infinite number of times.

IRC Internet Relay Chat: an online, real-time chat network distributed across the globe and accessible via the Internet.

Java A programming language invented at Sun Microsystems. It dramatically increases the interactive potential of the Web by allowing server machines to serve up small programs (called applets) that can be run on client machines without endangering security.

kick On IRC, a noun or verb referring to the expulsion of someone from a particular channel.

kill On IRC, a noun or verb referring to the extinguishing of someone's active connection to an IRC server.

lagged server On IRC, a server that for one reason or another is a step or two behind the rest of the network.

Lazarus A Usenet bot that monitored the alt.religion.scientology newsgroup for evidence of forged cancel messages.

mailbomb A large number of email messages sent simultaneously to one address, usually for aggressive purposes or retaliation.

mailbot A program that automatically filters email or performs other email upkeep functions.

mailing list An email discussion group, usually focusing on a particular topic, in which multiple email addresses receive every message anyone on the list sends to the list.

mailto button A piece of HTML code that creates an interactive button allowing email to be sent.

modbot On Usenet, an autonomous program that moderates a newsgroup by filtering out unwanted articles. Also called an *automoderator* or a *robo-moderator*.

MOO MUD object-oriented. MOOs belong to a subset of the world of multi-user domains (MUDs) and are distinguished by a programming structure that allows users to easily alter or extend the domain themselves.

MUD Multiuser domain, or multiuser dungeon. A MUD is a virtual world in which multiple Internet users interact in a shared space. MUDs began as text-based online versions of role-playing games like *Dungeons and Dragons* and have evolved into an array of forms that include 3-D animated environments.

natural language processing The AI discipline devoted to duplicating the processes of human speech.

netiquette The code of ethics and behavior that has independently evolved in cyberspace.

net split On IRC, a situation in which a computer or group of computers becomes separated from the rest of the network.

neural net   In AI, a computer program that attempts to duplicate neurological and biological systems.

news feed   On Usenet, the stream of articles that flows from server to server.

newsgroup   The basic unit of Usenet organization. The group of articles making up a newsgroup is ostensibly devoted to a particular topic.

nick collision   On IRC, a situation in which two entities (human or bot) have the same nickname at the same time in the same newsgroup. The result of a nick(name) collision is that both entities are kicked out of the channel.

operating privileges   On IRC, the right to determine the parameters of a specific channel. Also called *ops*.

process   A computer program in action.

protocol   A language that a computer uses to communicate with other computers.

robo-moderator   See *modbot*.

Robot Exclusion Protocol   On the Web, a standard for robot behavior aimed at protecting Web servers from robot abuses.

$alz convention   On Usenet, the labeling of a cancel message as such by including the word *cancel* in the header.

script   A simple programming shortcut consisting of a (usually) short sequence of instructions that automates a procedure.

search engine   A program that searches a database of information for items meeting specific criteria and then reports the results. The algorithms behind major Web search engines such as AltaVista, Yahoo!, and WebCrawler are closely guarded trade secrets.

server A computer or workstation that "serves" stored information or processing power to other machines, or "clients," on a network.

source code The code that makes up a computer program.

spam On Usenet, a noun or verb referring to the sending of one article to many recipients. Spam is more generally known as junk email.

spamdexing Including words on a Web page solely to trick search engine programs. For example, the word *sex* might be included a thousand times on a Web page in the hopes that anyone searching for sex would end up at that Web site.

spew An uncontrollable outpouring of Usenet articles into a newsgroup or multiple newsgroups.

spider See *Web robot*.

tag A basic building block of HTML code. The elements of HTML documents are marked by tags such as <BODY> and </BODY>, which indicate the start and end of the body of a document, and <I> and </I>, which indicate the start and end of italics.

telnet A protocol that allows one computer to connect to another on the Internet.

thread A sequence of messages related to a particular topic in an online discussion area or chat room.

URL Uniform resource locator: an address for a page on the Web or for other Internet resources.

Usenet The largest bulletin board computer conferencing system in the world. Thousands of newsgroups on this decentralized network are devoted to a seemingly infinite variety of topics.

warez Hacker slang for illegally copied programs.

Web robot A bot that travels from hyperlink to hyperlink, retrieving Web documents along the way and indexing information from them. Also called a *spider*.

wizard In MUDs, a person with special authority to make structural programming changes.

SOURCES

Some of the following people contributed as little as one email message to my research. Some contributed many hours of their time and wisdom. All of them made a difference.

Ryan Addams, Allan Alford, Zoran Alilovic, Nick Arnett, Jim Aspnes, Richard Bartle, Chris Behrens, Nathaneil Borenstein, Tim Bray, Seth Breidbart, Thomas Boutell, Lauren Burka Cancelmoose, Matt Carothers, Steve Chaney, Igor Chudov, Brian Clark, Alex Cohen, Ken Colby, Mary Conner, Peter da Silva, Michael Constant, Fernando Corbato, Joshua Cowan, David DeLaney, Dick Depew, Rahul Dhesi, Julian Dibbell, Abbe Don, Bruce Ellis, Samuel Epstein, Thomas Etter, etoy, Oren Etzioni, Lenny Foner, Joel Furr, Paul Ginsparg, Matthew Gray, Jonathan Hart, Rob Hartill, Dave Hayes, Eric Hillis, Susan Jacobson, Nick Jennings, Michael Johnson, Jonathan Kamens, Martijn Koster, Bruce Krulwich, David Lebling, John Leth-Nissen, John Levine, Greg Lindahl, Pattie Maes, Bill Mattocks, Michael Mauldin, Christine Maxwell, Doug McClaren, Jeanne McWhorter, John Milburn, Phillip Miller, Mandar Mirashi, Danny Mitchell, Kenrick Mock, Louis Monier, Cleo Odzer, Jarkko Oikarinen, Ove Ruben R Olsen, Rob Pike, Chris

Piepenbring, Brian Pinkerton, Robey Pointer, Michael Powers, Scott Reilly, Ed Rogers, Matt Rowley, Rich Salz, Ben Schneiderman, Ken Schweller, John Shepard, Stuart Shieber, Tim Skirvin, Peter Small, Homer Smith, Gene Spafford, Betsy Speicher, Robin Thelland, Tom Van Vleck, Joseph Weizenbaum, Thom Whalen, Bill Wisner, Jerry Yang, Scott Yelich, Marcy Yesowitch

SELECTED BIBLIOGRAPHY

Balabanovic, Marko, and Yoav Shoham. "Learning Information Retrieval Agents: Experiments with Automated Browsing." Paper presented at AAAI '95 (American Association for Artificial Intelligence) Spring Symposium on Information Gathering from Heterogeneous, Distributed Environments, Stanford University, Stanford, Calif., March 1995.

Ball, Gene, Dan Ling, David Kurlander, John Miller, David Pugh, Tim Skelly, Andy Stankosky, David Thiel, Maarten Van Dantzich, and Trace Wax. "Lifelike Computer Characters: The Persona Project at Microsoft Research." <www.research. microsoft.com/research/ui/persona/chapter/persona.htm> (16 May 1996).

Bester, Alfred. *The Stars My Destination*. New York: Vintage Books, 1956.

Brooks, Rodney A., and Pattie Maes, eds. *Artificial Life IV: Proceedings of the Fourth International Workshop on the Synthesis and Simulation of Living Systems*. Cambridge: MIT Press, 1994.

Chavez, Anthony, and Pattie Maes. "Kasbah: An Agent Marketplace for Buying and Selling Goods." Paper presented at

PAAM '96 (Practical Applications of Intelligent Agents and Multi-Agents), London, April 1996.

Cheong, Fah-Chun. *Internet Agents: Spiders, Wanderers, Brokers, and Bots.* Indianapolis: New Riders, 1996.

Chess, David. "Things That Go Bump in the Net." *Massively Distributed Systems.* <www.research.ibm.com/massdist/bump.html> (November 1995).

Chess, David, Benjamin Grosof, Colin Harrison, David Levine, Colin Parris, and Gene Tsudik. "Itinerant Agents for Mobile Computing." *Massively Distributed Systems.* <www.research.ibm.com/massdist/bump.html> (November 1995).

Chudov, Igor. *S.T.U.M.P. Robomoderator Program.* <www.galstar.com/~ichudov/usenet/scrm/robomod/robomod.html> (January 1997).

Cohen, Alexander J. "On the Origin of Artificial Life: Some Assembly Required." Paper presented at the Science and Literature Conference, Atlanta, October 1993.

Copeland, Jack. *Artificial Intelligence: A Philosophical Introduction.* Oxford: Blackwell, 1993.

Crevier, Daniel. *AI: The Tumultuous History of the Search for Artificial Intelligence.* New York: Basic Books, 1993.

Darwin, Charles. *The Origin of Species.* New York: Modern Library, 1993.

Dawkins, Richard. *The Blind Watchmaker: Why the Evidence of Evolution Reveals a Universe without Design.* New York: W. W. Norton, 1986.

Dery, Mark. *Escape Velocity: Cyberculture at the End of the Century.* New York: Grove Press, 1996.

Dewdney, A. K. *The Armchair Universe: An Exploration of Computer Worlds.* New York: W. H. Freeman, 1988.

Dreyfus, Hubert L. *What Computers Still Can't Do: A Critique of Artificial Reason.* Cambridge: MIT Press, 1992.

Edwards, Paul N. *The Closed World: Computers and the Politics of Discourse in Cold War America.* Cambridge: MIT Press, 1996.

Eichmann, David. "Ethical Web Agents." <rbse.jsc.nasa.gov/eichmann/www-f94/ethics/ethics.html> (April 1994).

etoy. *Digital Hijack.* <www.etoy.com/hijack-tank/over.html> (July 1996).

Etzioni, Oren. "Intelligence without Robots (A Reply to Brooks)." *AI Magazine,* December 1993.

Etzioni, Oren, and Daniel Weld. "A Softbot-Based Interface to the Internet." *Communications of the ACM* 37, no. 7 (July 1994).

Feigenbaum, Edward A., and Julian Feldman, eds. *Computers and Thought.* Menlo Park, Calif.: AAAI Press, 1995.

Flint, John, ed. *Riches from Another World.* Las Vegas, Nev.: Zon Association, 1996.

Foner, Leonard N. *What's an Agent, Anyway? A Sociological Case Study.* Agents Group, Agents Memo No. 93-01. Cambridge: MIT Media Laboratory, 1993.

Franklin, Stan, and Art Graesser. "Is it an agent or just a program?: A taxonomy for autonomous agents." In *Proceedings of the Third International Workshop on Agent Theories, Architectures, and Languages.* Springer-Verlag, 1996.

Gibson, William. *Neuromancer.* New York: Ace Books, 1984.

Ginsparg, Paul. "First Steps Towards Electronic Research Communication." *Computers in Physics* 8, no. 4 (July/August 1994): 390–396.

Gleick, James. *Chaos: Making a New Science.* New York: Penguin, 1987.

Gould, Stephen Jay. *Full House: The Spread of Excellence from Plato to Darwin.* New York: Harmony Books, 1996.

Gould, Stephen Jay. *Wonderful Life: The Burgess Shale and the Nature of History.* New York: W. W. Norton, 1989.

Graubard, Stephen R., ed. *The Artificial Intelligence Debate: False Starts, Real Foundations.* Cambridge: MIT Press, 1988.

Hafner, Katie, and Matthew Lyon. *Where Wizards Stay Up Late: The Origins of the Internet.* New York: Simon and Schuster, 1996.

Hafner, Katie, and John Markoff. *Cyberpunk: Outlaws and Hackers on the Computer Frontier.* New York: Simon and Schuster, 1991.

Haraway, Donna. *Simians, Cyborgs, and Women.* New York: Routledge, 1991.

Harrison, Colin G. "Smart Networks and Intelligent Agents." Paper presented at Mediacom '95, Southampton, UK, 11 April 1995.

Harrison, Colin G., David M. Chess, and Aaron Kershenbaum. "Mobile Agents: Are They a Good Idea?" *Massively Distributed Systems.* <research.ibm/massive/mobag.ps> (November 1995).

Harvey, David. *The Condition of Postmodernity.* Cambridge, Mass.: Blackwell, 1989.

Hauben, Michael, and Ronda Hauben. *Project: The Netizens and the Wonderful World of the Net: An Anthology Hypertext Version.* <woof.music.columbia.edu:80/~hauben/project_book.html> (April 1997).

Hayden, Charles. *Eliza Test.* <www.monmouth.com/~chayden/eliza/Eliza.html> (April 1997).

Holland, John H. *Hidden Order: How Adaptation Builds Complexity.* Reading, Mass.: Addison-Wesley, 1995.

Hutchens, Jason L. "How to Pass the Turing Test by Cheating." *alma*, no. 3. <www.diemme.it/~luigi/talk.html> (April 1997).

Johnson, Michael Boyle. "WAVESworld: A Testbed for Constructing 3D Semi-Autonomous Animated Computers." PhD diss., Massachusetts Institute of Technology, 1991.

Kelly, Kevin. *Out of Control: The New Biology of Machines, Social Systems, and the Economic World.* Reading, Mass.: Addison-Wesley, 1994.

Kidder, Tracy. *The Soul of a New Machine.* New York: Avon, 1981.

Koster, Martijn. "Robots in the Web: Threat or Treat?" *The Web Robots Pages.* <info.webcrawler.com/mak/projects/robots/threat-or-treat.html> (April 1997).

Koster, Martijn. "A Standard for Robot Exclusion." *The Web Robots Pages.* <info.webcrawler.com/mak/projects/robots/norobots.html> (April 1997).

Kroker, Arthur, and Marilouise Kroker. *Hacking the Future.* New York: St. Martin's, 1996.

Kroker, Arthur, and Michael A. Weinstein. *Data Trash: The Theory of the Virtual Class.* New York: St. Martin's, 1994.

Krulwich, Bruce. *Learning User Interests across Heterogeneous Document Databases.* Chicago: Andersen Consulting, 1995.

Langton, Christopher G., ed. *Artificial Life: An Overview.* Cambridge: MIT Press, 1995.

Lanier, Jaron. "Agents of Alienation." *Voyager.* <www.voyagerco. com/misc/jaron.html> (October 1996).

Lashkari, Yezdi, Max Metral, and Pattie Maes. "Collaborative Interface Agents." In *Proceedings of AAAI '94 Conference.* Seattle: AAAI Press, 1994.

Laurel, Brenda, ed. *The Art of Human-Computer Interface Design.* Reading, Mass.: Addison-Wesley, 1990.

Levy, Steven. *Artificial Life: A Report from the Frontier Where Computers Meet Biology.* New York: Vintage Books, 1992.

Levy, Steven. *Hackers: Heroes of the Computer Revolution.* New York: Dell, 1984.

Lewin, Roger. *Complexity: Life at the Edge of Chaos.* New York: Collier Books, 1992.

Lieberman, Henry. *Letizia: An Agent That Assists Web Browsing.* <lieber.www.media.mit.edu/people/lieber/Lieberary/Letizia/ Letizia-Intro.html> (16 November 1996).

Ma, Moses. *Avatars, Agents, and Bots, Oh My! A Classification for Presences in Multi-User Virtual Reality Simulations.* <206.79.196. 34/avatar/Avtdefn.html> (October 1996).

Maes, Pattie. "Agents That Reduce Work and Information Overload." *Communications of the ACM* 37, no. 7 (1994): 31–40.

Maes, Pattie. *The ALIVE System: Wireless, Full-Body Interaction with Autonomous Agents.* Perceptual Computing Technical Report No. 257. Cambridge: MIT Media Laboratory, 1995.

Maes, Pattie, and R. Kozierok. "Learning Interface Agents." In *Proceedings of AAAI '93 Conference.* Washington, D.C.: AAAI Press, 1993.

Mann, Steve. *Mediated Reality.* Perceptual Computing Technical Report No. 260. Cambridge: MIT Media Laboratory, 1995.

Mauldin, Michael L. "Chatterbots, Tinymuds, and the Turing Test: Entering the Loebner Prize Competition." In *Proceedings of AAAI '94 Conference.* Seattle: AAAI Press, 1994.

Mayr, Ernst. *One Long Argument: Charles Darwin and the Genesis of Modern Evolutionary Thought.* Cambridge: Harvard University Press, 1991.

Minsky, Marvin. *The Society of Mind.* New York: Touchstone, 1985.

Mumford, Lewis. *Technics and Civilization.* New York: Harcourt, Brace and World, 1934.

Negroponte, Nicholas. *Being Digital.* New York: Knopf, Vintage Books, 1995.

Pioch, Nicolas. *A Short IRC Primer.* <www.byz.org/~bi/docs/ircprimer.html> (April 1997).

Plato. *Collected Dialogues.* Edited by Edith Hamilton and Huntington Cairns. Princeton, N.J.: Princeton University Press, 1961.

Platt, Charles. "What's It Mean to Be Human Anyway?" *Wired,* April 1995.

Reeves, Byron, and Clifford Nass. *The Media Equation: How People Treat Computers, Television, and New Media Like Real People and Places.* New York: Cambridge University Press, 1996.

Reid, Elizabeth M. *Electropolis: Communication and Community on Internet Relay Chat.* <www.ee.mu.oz.au/papers/emr/electropolis.html> (April 1997).

Reilly, Scott. *Oz Project Home Page.* <www-cgi.cs.cmu.edu/afs/cs.cmu.edu/project/oz/web/oz.html> (October 1996).

Rheingold, Howard. *The Virtual Community: Homesteading on the Electronic Frontier.* Reading, Mass.: Addison-Wesley, 1993.

Shieber, Stuart. "Lessons from a Restricted Turing Test." *Communications of the ACM* 37, no. 6 (1994): 70–78.

Spafford, Eugene H. "Computer Viruses as Artificial Life." *Artificial Life* 1, no. 3 (Spring 1994): 249–265.

Starner, Thad. *The Cyborgs Are Coming, or the Real Personal*

Computer. <www-white.media.mit.edu/vismod/publications/tech_reports/abstracts/TR-318-ABSTRACT.html> (March 1996).

Starner, Thad, Steve Mann, Bradley Rhodes, Jennifer Healey, Kenneth B. Russell, Jeffrey Levine, and Alex Pentland. *Wearable Computing and Augmented Reality.* Vision and Modeling Group Technical Report No. 355. Cambridge: MIT Media Laboratory, 1995.

Sterling, Bruce. *The Hacker Crackdown: Law and Disorder on the Electronic Frontier.* New York: Bantam, 1992.

Stone, Allucquére Rosanne. *The War of Desire and Technology at the Close of the Mechanical Age.* Cambridge: MIT Press, 1995.

Tardo, Joseph, and Luis Valente. "Mobile Agent Security and Telescript." Paper presented at the IEEE (Institute of Electrical and Electronics Engineers) CompCon '96 Conference, Santa Clara, Calif., February 1996.

Thirunavukkarasu, Chelliah, Tim Finin, and James Mayfield. "Secret Agents: A Security Architecture for the KQML Agent Communication Language." Intelligent Information Agents Workshop, CIKM '95 (Conference on Information and Knowledge Management), Baltimore, December 1995.

Turing, A. M. "Computing Machinery and Intelligence." In *Computers and Thought,* edited by Edward A. Feigenbaum and Julian Feldman, 11–39. Menlo Park, Calif.: AAAI Press, 1963.

Turkle, Sherry. *Life on the Screen.* New York: Simon and Schuster, 1995.

Waldrop, M. Mitchell. *Complexity: The Emerging Science at the Edge of Order and Chaos.* New York: Simon and Schuster, Touchstone, 1992.

Weizenbaum, Joseph. *Computer Power and Human Reason.* New York: W. H. Freeman, 1976.

Williams, Joseph, ed. *Bots and Other Internet Beasties.* Indianapolis: Sams.net, 1996.

Winograd, Terry, and Fernando Flores. *Understanding Computers and Cognition: A New Foundation for Design.* Reading, Mass.: Addison-Wesley, 1986.

Wooldridge, Michael J., and Nicholas R. Jennings. "Agent Theories, Architectures, and Languages: A Survey." Paper presented at PAAM '96 (Practical Applications of Intelligent Agents and Multi-Agents), London, April 1996.

Zimmerman, Joy. "The Racter Factor." *Pacific Sun*, March 1–7, 1985, 2–6.

INDEX

FOR THE BEST IN PAPERBACKS, LOOK FOR THE

In every corner of the world, on every subject under the sun, Penguin represents quality and variety—the very best in publishing today.

For complete information about books available from Penguin—including Puffins, Penguin Classics, and Arkana—and how to order them, write to us at the appropriate address below. Please note that for copyright reasons the selection of books varies from country to country.

In the United Kingdom: Please write to *Dept. JC, Penguin Books Ltd, FREEPOST, West Drayton, Middlesex UB7 0BR.*

If you have any difficulty in obtaining a title, please send your order with the correct money, plus ten percent for postage and packaging, to *P.O. Box No. 11, West Drayton, Middlesex UB7 0BR*

In the United States: Please write to *Consumer Sales, Penguin USA, P.O. Box 999, Dept. 17109, Bergenfield, New Jersey 07621-0120.* VISA and MasterCard holders call 1-800-253-6476 to order all Penguin titles

In Canada: Please write to *Penguin Books Canada Ltd, 10 Alcorn Avenue, Suite 300, Toronto, Ontario M4V 3B2*

In Australia: Please write to *Penguin Books Australia Ltd, P.O. Box 257, Ringwood, Victoria 3134*

In New Zealand: Please write to *Penguin Books (NZ) Ltd, Private Bag 102902, North Shore Mail Centre, Auckland 10*

In India: Please write to *Penguin Books India Pvt Ltd, 706 Eros Apartments, 56 Nehru Place, New Delhi 110 019*

In the Netherlands: Please write to *Penguin Books Netherlands bv, Postbus 3507, NL-1001 AH Amsterdam*

In Germany: Please write to *Penguin Books Deutschland GmbH, Metzlerstrasse 26, 60594 Frankfurt am Main*

In Spain: Please write to *Penguin Books S. A., Bravo Murillo 19, 1° B, 28015 Madrid*

In Italy: Please write to *Penguin Italia s.r.l., Via Felice Casati 20, I-20124 Milano*

In France: Please write to *Penguin France S. A., 17 rue Lejeune, F–31000 Toulouse*

In Japan: Please write to *Penguin Books Japan, Ishikiribashi Building, 2–5–4, Suido, Bunkyo-ku, Tokyo 112*

In Greece: Please write to *Penguin Hellas Ltd, Dimocritou 3, GR–106 71 Athens*

In South Africa: Please write to *Longman Penguin Southern Africa (Pty) Ltd, Private Bag X08, Bertsham 2013*